Civil Engineering Structures According to the Eurocodes

Series Editor
Gilles Pijaudier-Cabot

Civil Engineering Structures According to the Eurocodes

Inspection and Maintenance

Xavier Lauzin

iSTE

WILEY

First published 2017 in Great Britain and the United States by ISTE Ltd and John Wiley & Sons, Inc.

ISTE Ltd
27-37 St George's Road
London SW19 4EU
UK

www.iste.co.uk

John Wiley & Sons, Inc.
111 River Street
Hoboken, NJ 07030
USA

www.wiley.com

Library of Congress Control Number: 2017939369

British Library Cataloguing-in-Publication Data
A CIP record for this book is available from the British Library
ISBN 978-1-78630-186-4

Contents

Introduction

An important factor in the design of new structures and repairs to existing structures is *the expected service life*.

In France, for major civil engineering works such as bridges, this duration is around 100 years (the British even go as far as 120 years).

For more modest structures such as water treatment plants, storage silos, etc., this duration was tacitly defined as around 50 years.

There are many factors that can influence this:

– the nature of the materials used in the construction (masonry, steel, concrete, wood, etc.);

– the quality of these materials (high-performance concrete, stainless steel, etc.);

– the constructive arrangements used (accumulation of water on metal structures, lack of encapsulation of steel in a reinforced concrete structure, etc.);

– quality of the execution (quality of the welding, implementation of concrete, etc.);

– monitoring and maintenance.

Within the context of the European Regulation for Calculation and Implementation, all these criteria have been taken into account when determining the duration of use of a structure.

This means that the design of the structures is obsolete if the maintenance conditions are not respected.

Let us recall section 2.4 of EN 1990:

"2.4 Durability

(1) The structure shall be designed such that deterioration over its design working life does not impair the performance of the structure below that intended, having due regard to its environment and the anticipated level of maintenance.

(2) In order to achieve an adequately durable structure, the following should be taken into account:

– the intended or foreseeable use of the structure;

– the required design criteria;

– the expected environmental conditions;

– the composition, properties and performance of the materials and products;

– the properties of the soil;

– the choice of the structural system;

– the shape of members and the structural detailing;

– the quality of workmanship, and the level of control;

– the particular protective measures;

– *the intended maintenance during the design working life*".

The question also arises for repairs carried out on a structure: what life expectancy should they be given?

With regard to new structures, EN 1990 indicates the following durations in Table I.1.

Design working life category	Indicative design working-life (years)	Examples
1	10	Temporary structures[a]
2	10–25	Replaceable structural elements, for example rolling beams, supporting devices
3	15–30	Agricultural structures and the like
4	50	Buildings and other structures
5	100	Monumental structures of buildings, bridges and other civil engineering structures

[a]Structures or parts of structures that can be disassembled for reuse should not be considered as temporary.

Table I.1. *Indicative design working life*

In the section "Execution of concrete structures", section 4.1 of EN 13670 also specifies the need for an inspection program:

> "(5) This standard assumes that the structure after completion is used as intended in the design and submitted to *planned inspection* and maintenance necessary to achieve the intended design working life and to *detect weaknesses or any unexpected behavior*".

This requirement implies providing access to the main structural elements at the design level.

Examples include:

– suspended bridges where replacement of the suspension has not been studied at design stage;

– water treatment plants for which it was not possible to empty the tanks (non-bypassable treatment line).

It also implies the need for a "*state 0*" during the reception for new constructions as well as a structure maintenance plan.

From this state, the sequence of tasks that is required to guarantee the duration of use of the structures is presented in the figure below:

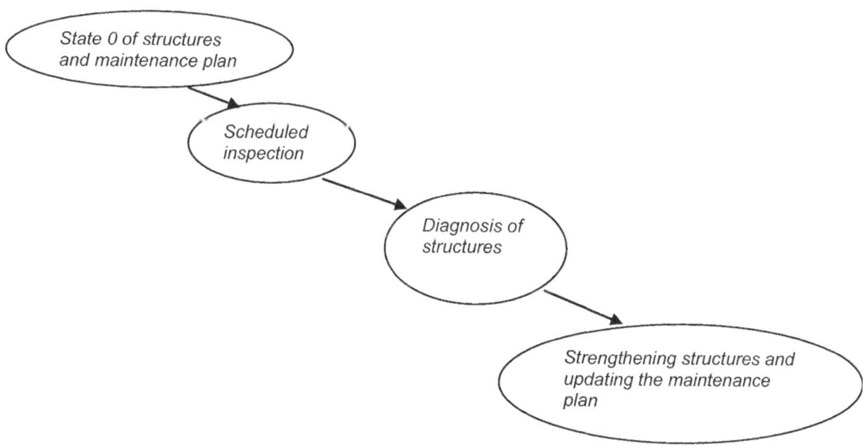

Figure I.1. *Sequence of tasks required to guarantee the duration*

The purpose of this book is to create an inventory of the methodologies used for inspections of civil engineering structures and to present the elements that can serve as a basis for the diagnosis and maintenance program of concrete structures.

We present the main topics that the reader can deepen their knowledge of by reading the standards cited.

How to use this guide

For a better understanding of the methodology used, in the last part of each inspection methodology listed in Chapter 1 is a paragraph about "points of to look out for", which refers to Chapter 3 for probable causes of the pathology and to Chapter 4 for the means of reinforcement that can be considered.

Chapter 2 gives the basic notions of resistance of the materials that are required for proper comprehension of the behavior of concrete and the interpretation of the observed disorders.

The examples in the Appendix are informative; they aim to show a type of connection in adequation with the inspected structure. They are purely formal.

1

Inspection of Structures: Methodologies

Inspection and diagnosis of structures are the most important phases of a maintenance operation and, eventually, of renovation. They require asking oneself a few questions before discussing the planning and recovery solution.

The questions are generally the following:

– what is the typology of the damage?

– what could be their cause?

– what is their scope?

– what is their probable evolution?

– what are the consequences for the structure?

– can the damage be repaired (technically and financially)?

To answer this question, the following methodology is usually applied:

– the first step involves a detailed visual assessment. This should be carried out by an expert civil engineer. It is similar to a health check without thorough analysis;

– the second step consists of a diagnosis by auscultation of the structure. This is managed by a civil engineer who relies on a specialized (and possibly multidisciplinary) laboratory.

1.1. Bridges

1.1.1. *General information*

For bridges, the Centre of Research and Expertise for Risks, Environment and Transport (CEREMA) formalized this principle and set up a methodology for monitoring and diagnosing this type of structure, which is summarized hereafter.

The management of structures is based on:

– *Recording of bridges:* this is a preliminary phase that consists of recognizing and recording the various heritage structures. The necessary data are: the type of structure, its *exact location*, its *main dimensional characteristics* and its *use*. The information should be verified in the field in order to take information into account that may not be included in the files;

– *The project file:* this is a document that gathers all the features of the structure along with its history. The contents of the file are defined in the Technical Instructions (ITSEOA);

– *Monitoring of structures:* this is of siginificant importance for maintaining the heritage and safety of users. It consists of following the evolution of various structures from a reference state (initial detailed inspection (IDI)), which is defined at the end of the construction or in the management takeover. The reference state can be modified by carrying out significant works such as expansion and extension. This monitoring is carried out over two levels:

- *periodic inspections*;

- *periodic detailed inspections*.

NOTE.– There is also a *detailed end-of-warranty inspection* to ensure the condition of a structure under contractual guarantee or 10-year liability.

1.1.1.1. *Periodic inspections*

Aim	Frequency	Requirements	Achievement
It applies to all structures if they are not carried out in the same year as another inspection (periodic or exceptional detailed inspection).	From 1 year (annual check) to 3 years (assessment visit) maximum.	– Detect any change in the pathologies that had already been noticed. – Take note of serious damages that pose a threat to users. – Identify the nature of routine or specialized maintenance.	Visual inspection without special access by trained agents.

Table 1.1. *Periodic inspections table*

1.1.1.2. *Periodic detailed inspections*

Aim	Frequency	Requirements	Achievement
Establish a health check of the structure and define the actions related to routine or specialized maintenance. It should be exhaustive and requires using means for access.	Six years but can be reduced to 3 years for weaker structures or increased to 9 years for robust structures. For underwater inspections, the frequency must be adapted according to the sensitivity of the structure (generally between 3 and 6 years).	– Check that the condition of the structure has not deteriorated abnormally. – Check that user safety devices are in good condition. – Check that there are no apparent threats to safety.	Visual inspection with special access carried out by agents who have received specific training.

Table 1.2. *Periodic detailed inspections table*

1.1.1.3. *Conditional monitoring actions*

These actions generally concern structures in exceptional conditions.

These are mainly as follows:

– exceptional visits or inspections following accidental events such as floods, landslides, violent storms, accidents, shocks, etc. or following observations from periodic inspections;

– enhanced monitoring or high-level monitoring activities for structures in critical condition.

Aim	Frequency	Requirements	Achievement
Complete the conventional monitoring actions and provide the information needed to carry out a major repair study (compilation of additional surveys, specific tests, sampling, etc.)	After examination during the periodic inspection, as a result of exceptional events, etc.	Establish a detailed diagnosis of the structure with a view to making major repairs.	Done by a specialized service provider with specific equipment.

Table 1.3. *Exceptional inspections table*

1.1.1.4. *Monitoring results*

The purpose of monitoring is to assess the level of service of a structure.

This service record can be classified as:

– *normal or quasi-normal*: structure generally in good condition (the only defects are due to routine maintenance) or minor defects that can be remedied by specific or specialized maintenance;

– *defective*: a structure with major structural damage for which the severity is assessed as likely to jeopardize the safety or durability of the whole structure;

– *doubtful*: analysis of a structure carried out at the end of a monitoring phase for which it was not possible to draw conclusions (actual or potential gravity, degradation of materials, etc.) or for which damages have not been highlighted (for example calcite sediment that may lead to corrosion of steel).

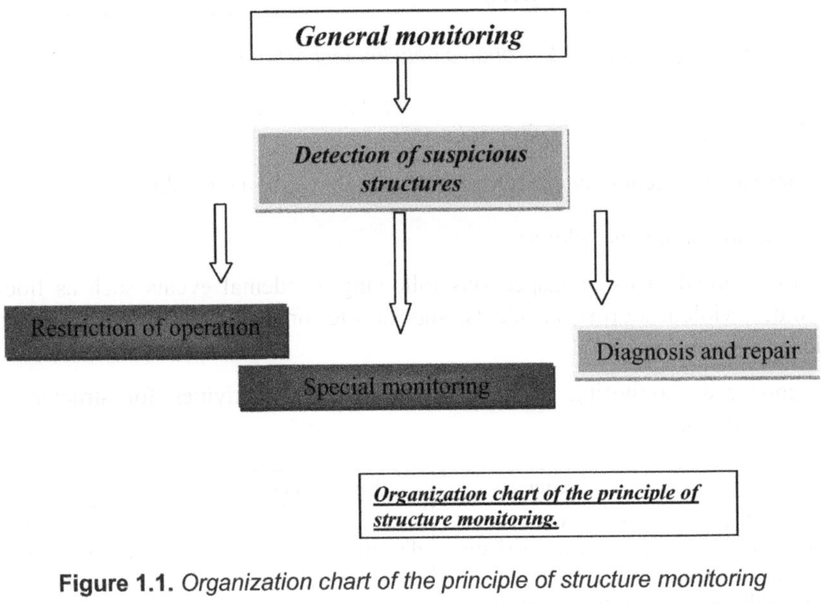

Figure 1.1. *Organization chart of the principle of structure monitoring*

1.1.2. *Regulatory documents*

1.1.2.1. *Booklet 3 of ITSEOA*

This booklet deals with "auscultation, enhanced monitoring, high-level surveillance, immediate safety measure or safeguard".

In particular, it defines:

– the approach to be followed in relation to the monitoring results;

– auscultation;

– enhanced monitoring;

– high-level monitoring;

– immediate safety and safeguarding measures.

1.1.2.2. *The revised ITSEOA from 1979*

This document includes the following structures:

– Booklet 01: Project files;

– Booklet 02: General information on monitoring;

– Booklet 03: Auscultation–enhanced monitoring–high-level monitoring security measures;

– Booklet 04: Topometric monitoring;

– Booklet 10: Aquatic foundations site;

– Booklet 11: Ground-site foundations;

– Booklet 12: Bearings;

– Booklet 13: Support devices;

– Booklet 20: Area of influence–access–approaches;

– Booklet 21. Equipment of structures (protection against water–coatings–road and sidewalk joints–railings–restraint systems);

– Booklet 30: Masonry bridges and viaducts;

– Booklet 31: Bridges made up of unreinforced and reinforced concrete;

– Booklet 32: Prestressed concrete bridges;

– Booklet 33: Metal bridges (steel, iron, cast iron);

– Booklet 34: Hanging bridges and cable-stayed bridges;

– Booklet 35: Emergency bridges;

– Booklet 40: Tunnels, covered trenches, protective galleries;

– Booklet 50: Metal nozzles;

– Booklet 51: Retaining structures;

– Booklet 52: Cuttings and embankments;

– Booklet 53: Protective structures.

1.1.3. *Human resources*

The achievement of a structure inspection service requires three levels of intervention:

– a project manager whose role it is to carry out the bid review, contract review, program review and file review. He is *the person in charge of the study*;

– a structure inspector whose role it is to intervene in each phase of the service in coordination with the project manager. He is *responsible for the report*;

– an inspection officer responsible for the inspection.

The qualification levels of various stakeholders are summarized in Table 1.4.

Function	Mission	Level
Project manager	Establish the diagnosis Propose a follow-up Finalize the inspection report	Bac+5 Bac+2
Inspector	Write the report Propose diagnostic elements	Bac+2 Bac
Inspection officer	Assist the inspector Carry out plans and monitoring	Bac

Table 1.4. *Qualification level table*

1.1.4. *Material resources*

A preliminary *preparation phase* is required to determine the material resources that are needed to carry out the inspection.

This phase is essential to ensure:

– stakeholder safety;

– quality of service.

In this context, the inspector will endeavor to verify:

– *visibility and accessibility* of the structures during a previsit with the site manager. He will thus be able to ascertain the presence of any vegetation, overhead lines, catenary lines, cleanliness and also identify any potential obstacles to carrying out the inspection;

– *means of access* to structures (propelled bridges or aerial platforms, vans equipped with collapsible scaffolding, ladders, ropes, craft, etc.).

Based on these elements, the intervention plan can be defined while bearing in mind the following elements:

– time required for technical and safety preparation;

– operational constraints of pathways leading to and from and crossed by the structure;

– delays in delivery of the service;

– the nature of structures to be inspected.

Before any intervention takes place, a risk analysis should be carried out, which should at least highlight the following points:

– definition of the conditions of intervention on frequented roads with the manager of the structure and preparation of requests for orders or notices for rerouting;

– verification of the conformity of means of access and staff qualification (CACES, etc.);

– EC certificate of the visiting craft;

– verification of PPE.

Inspections should always be carried out by two inspectors.

Each bridge inspector should have:

– a measuring tape, decameter and caliper;

– a digital camera;

– binoculars, magnifying glasses, flashlight;

– a fissurometer;

– a hammer, chisel, brush;

– a depth gauge;

– a spray can or a marker pen;

– a plumb line, spirit level;

– a bag for sampling;

– a rust scale;

– a measuring board;

– a mobile phone or walkie talkie.

1.1.5. *The project file*

Each structure has a file containing three subfolders that include the following:

– *Subfolder 1 "Design and construction"* contains all the information relating to the structure before it was put into service, in particular the Subsequent Intervention on the Structure File (SSIF);

– *Subfolder 2 "Reference state"* defines the initial state of the structure, which will serve as a reference for subsequent monitoring;

– *Subfolder 3 "Life of the structure"* contains the information after the reference date: VP of the monitoring actions, maintenance work, repairs, etc.

Elements necessary for the preparation of an inspection are as follows:

– for an IDI: the execution plans of the structure, calculation notes and technical sheets as well as a summary of the construction and repair checks;

– for an EDI: plans of the structure and reports of events;

– for a DEWI: the purpose and content of the guarantees;

– for all DIs, the Image Quality Structures (IQS) classification of the structure and previous inspection reports;

– the evolution of the level of operation (expansion, reloading of the rolling layers, limitation of loads, etc.);

– monitoring and auscultation VP (topo, cracks, thickness, etc.).

It is also important to have:

– the date of construction (for understanding the constructive dispositions, recalculation of the structure, etc.);

– the method and phasing of the project;

– the materials and processes used;

– the foundation method;

– possible on-site incidents.

1.1.6. *How an inspection is carried out*

The role of the inspector in the execution of the inspection program is to:

– establish access means and equipment;

– evaluate meteorological conditions (rain, wind, snow, ice, etc.) with indications of temperatures;

– get a record of special conditions;

– conduct a close visual inspection detailing any defects encountered. Any defect shall be characterized by:

- its type (crack, spalling, etc.);
- its physical appearance and dimensions;
- its extent;
- its location.

The observations to be made on-site and to be recorded in the inspection report include:

– the area of influence (embankment, excavation, environment, etc.);

– the deck (extrados and intrados);

– the equipment (roadway, sidewalks, storm water system, cornices, guardrails, gates, waterproofing, road joints, monitoring devices, etc.);

– the support system (bosses, bearings, etc.);

– supports (piers, abutments);

– the foundations (on land, river or maritime sites, protection against shocks, etc.);

– accessibility;

– crossings (nature of the roadway crossed, nature of the crossing, clear height, crossing gauge, etc.);

– the characteristics of the structure.

1.1.7. *The inspection report*

The inspection report must include:

– a chapter identifying the structure;

– a chapter specifying the general characteristics;

– a chapter containing information on the design and execution of the structure;

– a chapter on the life of the structure;

– a chapter on the findings and measurements carried out as part of the inspection;

– a chapter on tests, auscultations, investigations;

– a summary chapter on the state of the structure and its evolution;

– appendices with:

  - plans of the structure (longitudinal, transverse, elevation);
  - plans and diagrams of pathologies encountered;
  - photographic report.

During evaluation visits (IQS visits), the classification of structures is shown in Figure 1.2.

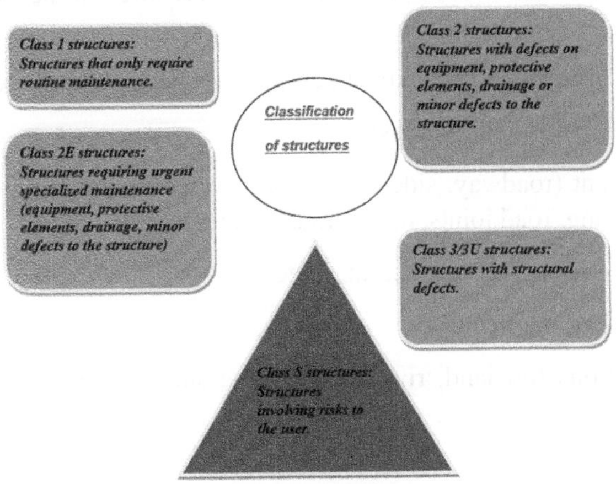

Figure 1.2. *Classification of structures*

1.1.8. *Points to look out for*

Points to keep an eye on are listed by type of structure in each IQS booklet from the CEREMA database.

1.1.9. *Classification example*

Observations	Class	Photos
North: – Efflorescences due to trails of lime contained in the concrete through the internal water circulation and its deposit in the form of calcite in the cladding.	1	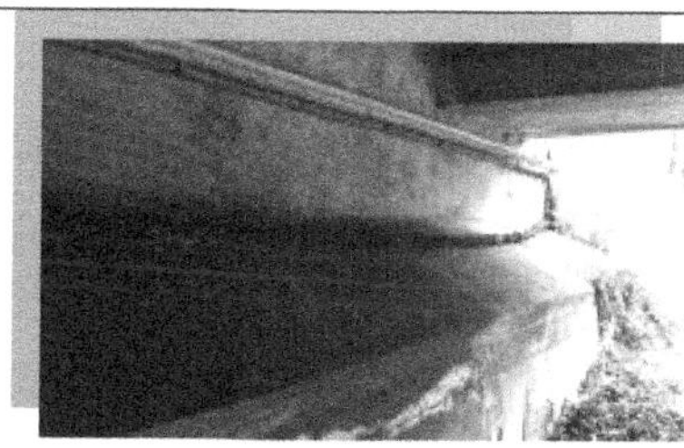
South: – Bursting and loosening of the patching concrete due to the thrust exerted by oxidation of the reinforcements and accumulation of water between the spacer and the sole of the abutment.	2	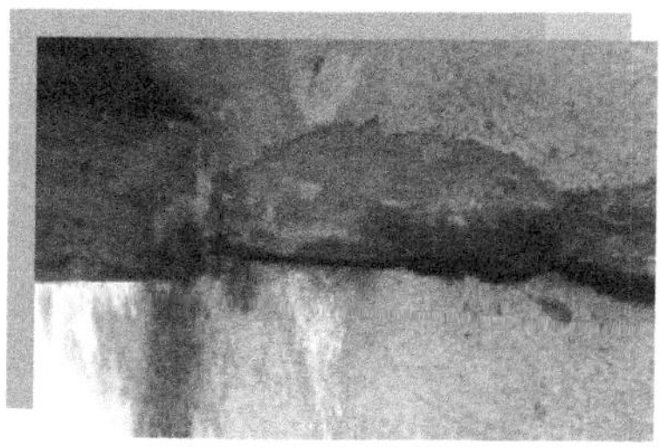

1.2. Structures for the retention and transportation of liquids

1.2.1. *General information*

In the same vein as for structures, the CEMAGREF published a guide to the ITBTP editions (ITBTP annals no. 532 of March/April 1995) under the title of:

> "Pathology and repair of concrete structures for the storage and transportation of liquids".

The CEMAGREF guide has two parts:

– the first part looks at the possible modes of repair depending on the type of damage and the severity index;

– the second part gives a detailed review of the inspection methodology, the pathologies and the choice of repair techniques.

It specifies the elements necessary for carrying out a diagnostic test of liquid retention structures, in particular the following points:

– knowledge of the structure's history;

– quantitative and qualitative description of the various damage;

– identification and extent of the various damage;

– recognition of the physicochemical characteristics of the base material;

– comparison of these characteristics in healthy areas and altered areas;

– the parameters test determining the main pathologies that are generally recognized on the type of structure being considered;

– an assessment of the likely evolution of the damage;

– if necessary, recalculation of the structure (reinforcement).

Along the same lines as the CEREMA guide for civil engineering structures, the CEMAGREF guide proposes the following methodology for evaluating structures for storage and transportation of liquids.

Steps	Type of investigation
1	*Inventory of structures to be inspected.* *Examination of the project file.* *Summary inspection and initial evaluation* in the normal operation of the structure, usually dedicated to the owner.
2	*Detailed inspection* of the structure. *Complementary investigations.*
3	*Detailed civil engineering inspection* concerning the quantification and qualification of the damage that affects the structure; this could be accompanied or not by *a diagnosis* of the materials and the structural behavior by auscultation and/or instrumentation.
4	*A diagnosis* to bring the structure back to its initial operating objectives or to a higher level of service (reinforcement) or demolition.

Table 1.5. *Methodology for evaluating structures*

1.2.1.1. *Step 1*

For this stage, the report should provide the following information:

– location, type of environment and information specific to this type of structure;

– general characteristics of the structures (constituent materials, type of foundation, roofing, waterproofing, etc.);

– technical and dimensional characteristics of the structures (studying the "project file", which includes formwork and reinforcement plans, calculation notes and technical details such as the treatment of the concreting reworks, etc.);

– the type of internal waterproofing selected at the design stage and carried out on the site;

– the type of external waterproofing of the structure (roof, buried part of structures, etc.);

– previous maintenance and maintenance procedures;

– visual inspection accompanied by a photograph file. The photos should be listed and localized;

– an initial evaluation of structures according to the codification below.

1.2.1.2. *Step 2*

If the report in step 1 classifies the structure as levels 2, 3 or 4 in the Table 1.7, a more detailed inspection of the structures must be carried out and additional investigations can be considered:

determination of the physical and chemical characteristics of concrete and other materials (waterproofing, etc.). Core drilling of concrete structures is usually carried out on structures that compression tests and chemical characterization tests are carried out on;

– determination of the characteristics of steel coating (for example pachometric tests);

– instrumentation and monitoring of identified pathologies.

A complementary report will then be produced by analyzing the evolution of pathologies, repairs that can be considered, the constraints on the operation and maintenance of structures.

1.2.1.3. *Step 3*

This is the proper diagnostic phase, which encompasses the set of steps 1–3.

It is carried out by a civil engineering expert and must reveal the following points:

– determination of the causes of pathologies;

– evaluation of the structure overall and per component;

– indication of repair or demolition solutions with the technical requirements inherent to the different processes;

– recalculation of structures;

– evaluation of the cost of repairs;

– estimated service life after repair.

1.2.1.4. *Step 4*

This is the project of renovating a structure once the repair solution is chosen.

1.2.2. *Regulatory documents*

The aforementioned CEMAGREF guide; it may be supplemented by the CEREMA guides for civil engineering works.

1.2.3. *Human resources*

An inspection service involves three levels of intervention:

– a civil engineering inspector whose role it is to intervene in each phase of the service in coordination with the inspection officer. He is *responsible for the report*;

– an inspection officer who is responsible for inspection.

The qualification levels of the various stakeholders are summarized in Table 1.6.

Function	Mission	Level
Civil engineering inspector. Project manager	Establish the diagnosis Propose follow-up Finalize inspection report	Bac+5 Bac+2
Inspection officer	Assist the inspector Carry out plans and monitoring	Bac

Table 1.6. *Stakeholder qualification levels*

1.2.4. *The material means*

The determination of the material means necessary for carrying out the inspection requires a preliminary *preparation* phase.

This phase is essential to ensure:

– stakeholder safety;

– quality of service.

In this context, the inspector will endeavor to verify:

– *visibility and accessibility* of the structures during a previsit with the manager. He will thus be able to ascertain the presence of any vegetation, overhead lines, catenary lines, cleanliness and identify any potential obstacles to carrying out the inspection;

– *means of access* to structures (propelled bridges or aerial platforms, vans equipped with collapsible scaffolding, ladders, ropes, craft, etc.).

Based on these elements, the intervention plan can be defined while bearing in mind the following elements:

– time required for technical and safety preparation;

– operational constraints of the structures (draining, cleaning, etc.);

– delays in delivery of the service (for example inspection of the tank of a drinking water reservoir during the period of cleaning and disinfection);

the nature of structures to be inspected.

Before any intervention takes place, a risk analysis should be carried out, which should at least highlight the following points:

– definition of the conditions of intervention with the manager, in particular if access and inspection requires work on ropes;

– verification of the conformity of means of access and qualification of staff (CACES, etc.);

– EC certificate of the visiting craft;

– verification of sensors (CH_4, H_2S, etc.);

– verification of PPE.

Like for bridges, an inspection should generally be carried out by two inspectors.

Each inspector should have:

– a measuring tape, decameter and caliper;

– a digital camera;

– binoculars, magnifying glasses, flashlight;

– a fissurometer;

– a hammer, chisel, brush;

– a depth gauge;

– a spray can or a marker pen;

– a plumb line, spirit level;

– a bag for sampling;

– a rust scale;

– a measuring board;

– a mobile phone or walkie talkie.

1.2.5. *The project file*

Each structure has a file containing three subfolders that include the following:

– *Subfolder 1 "Design and construction"* contains all the information relating to the structure before it is put into service, in particular the SSIF;

– *Subfolder 2 "Reference state"* defines the initial state of the structure, which will serve as a reference for subsequent monitoring;

– *Subfolder 3 "Life of the structure"* contains the information after the reference date: VP of the monitoring actions, maintenance work, repairs, etc.

1.2.6. *How the inspection is carried out*

The role of the inspector in the execution of the inspection program is to:

– establish access means and equipment;

– evaluate meteorological conditions (rain, wind, snow, ice, etc.) with indications of temperatures;

– get a record of special conditions;

– conduct a close visual inspection detailing any defects encountered. Any defect shall be characterized by:

- type (fissure, spalling, etc.);

- physical appearance and dimensions;

- the extent;

- location.

The observations to be made on-site and to be recorded in the inspection report include:

– the area of influence (environment of underground area, aerial zone, etc.);

– the tank (area in contact with the liquid, area in contact with air, area in contact with the ground);

– equipment (guard rails, ladders, cover seal, interior waterproofing of the basin, etc.);

– tower tank columns;

– foundations (on land, river or maritime sites, protection against shocks, etc.);

– surroundings and access;

– the features of the structure.

1.2.7. *The Inspection report*

The inspection report must include:

– a chapter identifying the structure;

– a chapter specifying the general characteristics;

– a chapter containing information on the design and execution of the structure;

– a chapter on the life of the structure;

– a chapter on the findings and measurements carried out as part of the inspection;

– a chapter on tests, auscultations, investigations;

– a summary chapter on the state of the structure and its evolution;

– appendices with:

- plans of the structure (longitudinal, transverse, elevation);
- plans and diagrams of pathologies encountered;
- photographic report.

The classification of damages proposed in this methodology can be summarized as seen below.

Level	Defects class	Description of the level	Follow-up (type of investigation)
1	A	Structure in good condition (new or old, without defects).	Nothing in particular to report, follow-up and normal maintenance of the structure (annual, biannual depending on nature) Periodic inspection.
	B	Defects existing right from the beginning of the structure and with no significant consequence other than aesthetic.	
2	C	Some defects, risk of abnormal evolution.	Visual inspection.
3	D	D1: defects that show some evolution D2: defects that indicate advanced development for parts that are not in contact with liquids D3: defects that show an advanced evolution for parts in contact with liquids.	Detailed civil engineering inspection possibly with tests on materials.
	E	Defects that reflect a change in the structural behavior of the structure involving its life expectancy (or use).	
4	F	The structure cannot function reliably. The risk of ruin is significant. Possible first-aid solutions and/or demolition of the structure must be considred.	Complete and instrumented diagnosis of the structure with auscultation and sampling.

Table 1.7. *Classification of damages table*

An example of classification of a tower tank can be seen in the table below.

Parts of the structure	Definition of probable causes	Severity index	Possible repair solution
Surroundings		B	
Tank support posts	Carbonation of concrete	B	Technical painting after purging and local repairs
Dome	Carbonation of concrete and corrosion of steel	C	Sealing of the tank
Tank and tank walls (interior)	Carbonation of concrete and corrosion of steel	C	Sealing of the tank
Tank walls (exterior)	Carbonation of concrete and corrosion of steel	C	Technical painting after purging and local repair
Cover dome	Carbonation of concrete and corrosion of steel	E	Recover the subsurface with shotcrete after purging. Additional protection against moisture.
Overall structure (max. severity index)		E	

NOTE.– A full example can be found in Appendix 1.

1.2.8. *Points to look out for*

The tables below list the points to keep an eye on and links them with a severity index and repair solutions.

1.2.8.1. *Concrete structures*

Type of defect	Probables causes	Severity index	Repair solution
Steel portion	Cover defect Shock (see Chapter 3, section 3.1)	D or E	Timely repair (see Chapter 4, section 4.1.5) Shotcrete (see Chapter 4, section 4.1.4) Cathodic protection (see Chapter 4, section 4.2)
Segregation	Sealing failure of formwork. Poor implementation of concrete. Inadequate rheology.	D	Timely repair (see Chapter 4, section 4.1.5)

Bubbling	Poor implementation of concrete. Inadequate rheology.	On raw concrete structure: B On waterproof or adherent waterproofing support structure: D	Timely repair (see Chapter 4, section 4.1.5)
Disintegration of concrete	Quality of concrete not adapted to the environment Implementation defect Abrasion from sand carried by water. (see Chapter 3, section 3.1)	D	Depending on the results of the chemical analysis of the concrete (see Chapter 4) In the latter case, an antiabrasion mortar may be considered.
Concrete peeling	Shock Aggressive environment Quality of concrete (see Chapter 3, section 3.1)		Timely repair (see Chapter 4, section 4.1.5) Shotcrete (see Chapter 4, section 4.1.4)
Faience	Withdrawal Alkali-reaction Internal sulfate reaction (see Chapter 3, section 3.1)	B D, E or F D, E or F	Removal: protection by technical painting (I3, I4) In the case of alkali-reaction or ISR to be seen depending on chemical analyses
Isolated cracks $w \leq 2/10$ mm $2/10 < w \leq 20/10$ $20/10 < w$	(see Chapter 3, section 3.1) Check whether the crack has changed		Timely repair (see Chapter 4, section 4.1.5) in first two cases To analyze in the third case (lizards)
Multiple cracks $w \leq 2/10$ mm $2/10 < w \leq 20/10$ $20/10 < w$	(see Chapter 3, section 3.1) Check whether the crack has changed		Timely repair (see Chapter 4, section 4.1.5) Shotcrete (see Chapter 4, section 4.1.4) Additional prestressing (see Chapter 4, section 4.1.3)
Cracks from loading or unloading	Sizing defect (dynamic effects) (see Chapter 3, section 3.1; check whether or not the crack has changed)		Shotcrete (see Chapter 4, section 4.1.4) Additional prestressing (see Chapter 4, section 4.1.3)

Visible foundations	Scouring, compaction of the soil around the silo	E or F	Backfill. Recovery in the underground (see Chapter 4, section 4.3)
Verticality defect	Differential settlement Hydrology of the site Compaction of the embankment Evacuation of storm water	Rotation without influence on operation: D Rotation that does not compromise stability and waterproofness: E Rotation that compromises stability and/or operation: F	Recovery in the underground (see Chapter 4, section 4.3)

1.2.8.2. *Masonry structures*

Type of defect	Probables causes	Severity index	Repair solution
Alteration of masonry	Environmental aggression	Superficial B or C In the mass D or E	According to the chemical analysis of the pathogen (remineralizing, coating, etc.)
Crumbling masonry	Mechanical or chemical aggression	C or D	Reconstitution or replacement
Shattering	Compression, freezeshock, etc.	C or D	Reconstitution or replacement
Grouting defect	Chemical alteration	Localized D to E Widespread E to F	Timely restoration (see Chapter 4, section 4.1.5) Projected mortar (see Chapter 4, section 4.1.4)
Corrosion of tie rods	Corrosion	C to E	Treatment or replacement. Cathodic protection (see Chapter 4, section 4.2)
Cracking of masonry	see Chapter 3, section 3.1	C to E	Timely restoration (see Chapter 4, section 4.1.5) Projected mortar (see Chapter 4, section 4.1.4) Recovery in the underground (see Chapter 4, section 4.3)

1.2.8.3. *Exterior coatings*

Type of defect	Probables causes	Severity index	Repair solution
Peeling off	Adhesion defect Defective humidification of the substrate. Freeze (see Chapter 3, section 3.1)	B or C	Partial detachment: localized recovery possible General detachment: total repairs after demolition
Faience	Withdrawal. Defective humidification of the substrate. (see Chapter 3, section 3.1)	B or C	Protection by technical paints
Chalking	Drying defect. Defective humidification of the substrate (see Chapter 3, section 3.1)	B or C	Protection by technical paints
Cracking	See Chapter 3, section 3.1	B–D	Depending on the nature of the cracking

1.2.8.4. *Waterproofing coatings based on hydraulic binders*

Type of defect	Probables causes	Severity index	Repair solution
Peeling off	Adhesion defect Defective humidification of the substrate. Freeze Underground pressure (see Chapter 3, section 3.1)	C or D	Partial detachment: localized recovery possible General detachment: total repairs after demolition.
Faience	Withdrawal. Defective humidification of the substrate. (see Chapter 3, section 3.1)	B or C	Protection by technical paints.
Chalking	Drying defect. Defective humidification of the substrate (see Chapter 3, section 3.1)	B or C	Protection by technical paints.
Cracking	see Chapter 3, section 3.1	B–D	Depending on the nature of the cracking.

1.2.8.5. *Tank waterproofing coatings based on synthetic resins*

Type of defect	Probables causes	Severity index	Repair solution
Peeling off	Adhesion defect Underground pressure (see Chapter 3, section 3.1)	Localized defects: C Widespread defects: D	Partial detachment: localized recovery possible General dislocation: total repairs after demolition.
Polymerization defect	Poor composition Poor implementation Commissioning was too fast (see Chapter 3, section 3.1)	D	Total rebuild after demolition.
Chalking	Physicochemical attack (UV type) Product evolution (see Chapter 3, section 3.1)	C	Total rebuild after demolition.
Cracking	See Chapter 3, section 3.1	D	Depending on the nature of the cracking.

1.2.8.6. *Waterproofing membranes*

Type of defect	Probables causes	Severity index	Repair solution
Peeling off Blistering	Adhesion defect Underground pressure (see Chapter 3, section 3.1)	Localized defects: C Widespread defects: D	Complete replenishment or replacement of the membrane
Sealing defect at the welds	Poor welding Poor material	D	Recovery of seals
Sealing defect at singular points	Complexity of welding	D	Resumption of singular points (resin, etc.)
Surface degradation	Physical–chemical attacks	D	Complete replenishment or replacement of the membrane

1.3. Storage structures for petroleum products

1.3.1. *General information*

The topic discussed here is mainly that of concrete structures located inside a fuel park.

It therefore concerns the retention basins and the foundations of tanks, the latter mostly being made up of steel in accordance with the CODRES requirements and are the subject of DT 94.

Figure 1.3. *Retention basin of an oil storage tank*

The UFIP (*Union Française des Industries Pétrolières* – the French Petrolium Industries Union) has produced a guide:

"Maintenance guide for civil engineering works and structures" (DT 92).

This guide provides instructions for the setting up of a monitoring procedure by field agents instructions comprising:

– monitoring visits;

– visits with increased control.

The inspected structures (mainly retention basins and tank foundations) are classified according to the level of danger of the products stored within it.

Type of structure	Classification
All structures except those in category II	Category I
Critical structures in terms of environmental risk (see "Professional guide for defining the perimeter as part of the modernization plan") Storage of flammable liquids.	Category II

Table 1.8. *Classification of structures according to the level of danger*

Along the same lines as for bridges, the guide recommends creating a "monitoring record" that contains the following elements:

– a summary technical sheet specifying the location of the structure and its description, the geometric and technical characteristics, its category (I or II);

– a technical file containing the project documentation (formwork and reinforcement plan, calculation notes, type of waterproofing, type of foundation, geotechnical studies, etc.), a history of the interventions carried out on the structure (structural modifications, replacement of pipes, change of fire seals, etc.), inspections already carried out.

This monitoring file must be accessible at each periodic inspection and updated after each inspection.

A *monitoring program* shall then be established including:

– the classification of the structure according to its condition:

Condition of structure	Definition of the category	Comments	Nature of the intervention
1	Satisfactory condition only requiring routine maintenance		Cleaning of basins and expansion joints Cleaning of drainage Control of access devices to basins, pipes, etc.
2	Fair condition with mild damage that is beyond routine maintenance	Specialized maintenance should be provided.	Drainage repair Recovery of expansion joints Repair of local damage (small cracks, spalling, etc.) Treatment of corrosion of metallic elements Repair of sealants and fire protection provisions
2E	The same as state 2 but with a risk of evolution of the pathologies (evolutionary state)	Implementation of enhanced monitoring	
3	Degraded structural condition requiring repair work	Diagnosis and repair	Major structural repairs (walls, paving, foundations, etc.) Replacing anchor bolts Installation of a structure instrumentation
3P	The same as state 3 but with a priority deadline (integrity, retention capacity, bearing capacity that can be quickly defected)	Diagnosis and repair as soon as possible	

– the frequency of monitoring visits should be dependent on the state of conservation and the category of the structure, in other words at least:

- 1 year for Category II structures;

- 5 years for Category I structures.

1.3.2. *How the inspection is carried out*

The previous requirements for bridges or tanks are also valid for oil repositories with the following specificities:

– as the installation is classified with respect to the environmental risks, each visit requires an application for authorization;

– the visiting equipment must include at least the individual safety equipment, a camera that meets the ATEX zone requirements, measuring tools, etc.

1.3.2.1. *Periodic visits*

At the end of the visit, the agent should draw up a monitoring card comprising the following points:

– the nature of the structure and its category;

– its location or denomination;

– a precise description of the pathologies;

– the level of damage (D1, D2, D3) according to the above classification;

– the results from a counter-visit;

– the need to re-evaluate the structure after further analysis and investigation.

Following an analysis of the monitoring sheets or additional investigation (if these are necessary), the final classification of structures is carried out as follows:

– *a structure is class 1* if no level 2 or 3 damage has been noted on any of the components;

– *a structure is class 2* if no level 3 damage has been found on any of the components but if there is at least one level 2 damage;

– *a structure is class 3* (or 3P) if a level 3 (or 3P) damage has been detected on at least one of the components.

1.3.2.2. *Visits with reinforced control*

To assess the risk of evolution of the damage, an action plan will define the details of the checks to be carried out, such as:

– evolution of a crack opening or a cracked surface;

– verification of the verticality of a storage basin;

– control of foundation compaction.

The reinforced control must conclude either to the absence of evolution risk or to the need for repair.

The response times are summarized in Table 1.9.

Final classification of the structure	Actions to be taken	Implementation timeframes
1		
2E	Reinforced control	According to action plan
2	Corrective operations (according to action plan)	5 years maximum or during the deactivation of the reservoir (*) if it occurs within 5 years
3	Corrective operations (according to action plan)	3 years maximum or during the deactivation of the reservoir (*) if it occurs within 3 years
3P	Implementation of priority measures	6 months maximum
	Corrective operations (according to action plan)	3 years maximum or during the deactivation of the reservoir if it occurs within 3 years

NOTE.– A full example is given in Appendix 2.

Table 1.9. *Periodicity table according to DT92*

1.3.3. *Specificities for this type of structure*

The different basins (according to INRS) are presented in Figures 1.4–1.6.

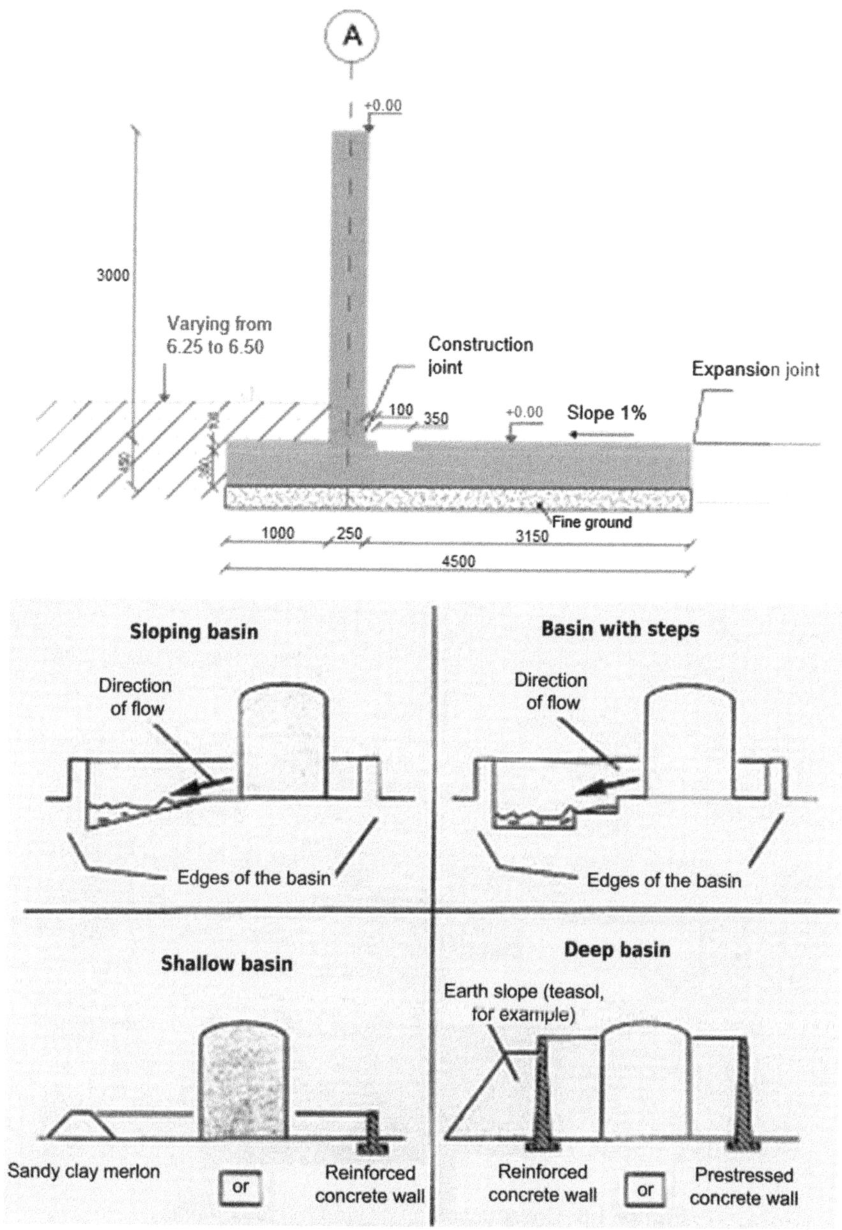

Figure 1.4. *Detailed diagram of the retention basin wall*

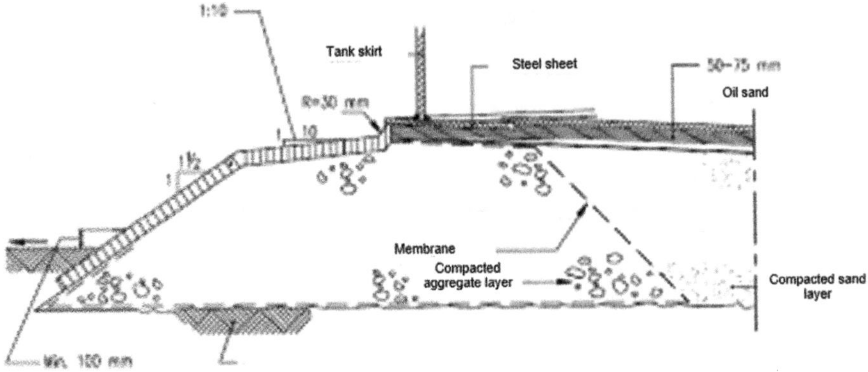

Figure 1.5. *Detailed diagram of the tray of the basin bottom*

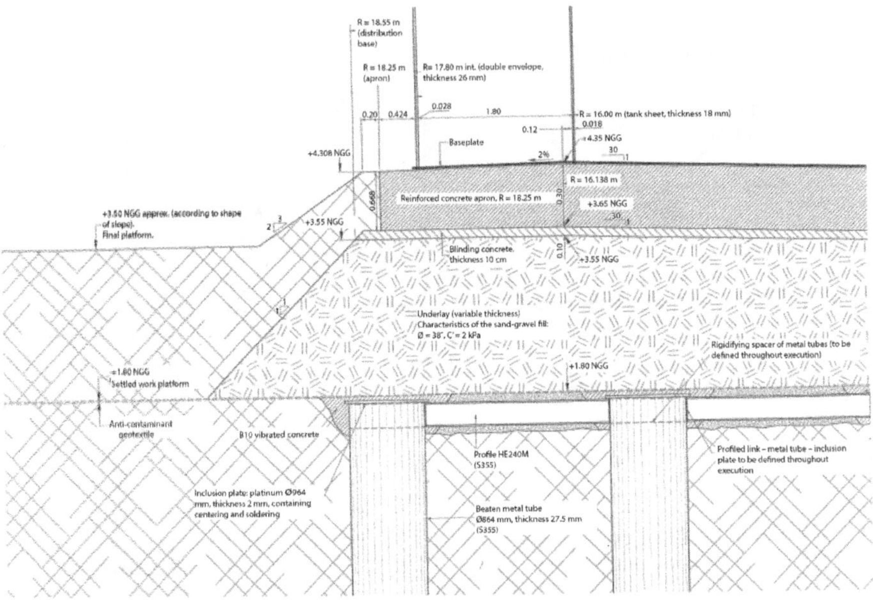

Figure 1.6. *Construction on ground reinforcements*

1.3.4. *Points to look out for*

1.3.4.1. *Concrete base foundation (peripheral base)*

Type of defect	Probables causes	Severity index	Repair solution
Settlement Sinking	Geotechnical study defect Scouring	2E to 3	Recovery in the underground (see Chapter 4, section 4.3)
Cracking	Mechanical malfunction. Physicochemical alteration (alkali-reaction, ISR, etc.) see Chapter 3, section 3.1	3–3P	Dependent on the nature of the pathology
Concrete degradation	Physicochemical alteration (alkali-reaction, ISR, etc.) see Chapter 3, section 3.1	2–3	Timely repairs (see Chapter 4, section 4.1.5) Projected mortar (see Chapter 4, section 4.1.4)
Visible steel	Cover defect Shock (see Chapter 3, section 3.1)	2–3	Timely repairs (see Chapter 4, section 4.1.5) Projected concrete (see Chapter 4, section 4.1.4) Cathodic protection (see Chapter 4, section 4.2, with particular attention for ATEX areas)
Corroded or absent anchor bolts	Aggressive environment	2–3	Replacement

1.3.4.2. *Foundations of the basin on soft foundation base*

Type of defect	Probables causes	Severity index	Repair solution
Settlement sinking	Geotechnical study defect Scouring	2E to 3	Recovery in the underground (see Chapter 4, section 4.3)
Leak detection drain defects	Shear break	3P	Replacement

1.3.4.3. *Concrete structures (low walls, blocks, etc.)*

Type of defect	Probables causes	Severity index	Repair solution
Settlement sinking	Geotechnical study defect Scouring	2E to 3	Recovery in the underground (see Chapter 4, section 4.3)
Cracking	Mechanical malfunction. Physicochemical alteration (alkali-reaction, ISR, etc.) see Chapter 3, section 3.1	3–3P	Dependent on the nature of the pathology
Concrete degradation	Physicochemical alteration (alkali-reaction, ISR, etc.) see Chapter 3, section 3.1	2–3	Timely repairs (see Chapter 4, section 4.1.5) Projected mortar (see Chapter 4, section 4.1.4)
Visible steel	Cover defect Shock (see Chapter 3, section 3.1)	2–3	Timely repairs (see Chapter 4, section 4.1.5) Projected concrete (see Chapter 4, section 4.1.4) Cathodic protection (see Chapter 4, section 4.2, with particular attention for ATEX areas)
Degraded seals	Wear, environmental conditions, etc.	3–3P	Replacement

1.3.4.4. *Paving with sealing function*

Type of defect	Probables causes	Severity index	Repair solution
Settlement sinking	Geotechnical study defect Scouring	2E to 3	Recovery in the underground (see Chapter 4, section 4.3)
Cracking	Mechanical malfunction. Physicochemical alteration (alkali-reaction, ISR, etc.) see Chapter 3, section 3.1	3–3P	Dependent on the nature of the pathology
Concrete degradation	Physicochemical alteration (alkali-reaction, ISR, etc.) see Chapter 3, section 3.1	2–3	Timely repairs (see Chapter 4, section 4.1.5) Casting of a new slab Implementation of protective resin
Visible steel	Cover defect Shock (see Chapter 3, section 3.1)	2–3	Timely repairs (see Chapter 4, section 4.1.5) Projected concrete (see Chapter 4, section 4.1.4) Cathodic protection (see Chapter 4, section 4.2, with particular attention for ATEX areas)
Degraded seals	Wear, environmental conditions, etc.	3–3P	Replacement

1.3.4.5. *Bottom of basin made of earth*

Type of defect	Probables causes	Severity index	Repair solution
Settlement sinking	Geotechnical study defect Scouring	2E–3	Recovery in the underground (see Chapter 4, section 4.3)
Vegetation animals	Lack of maintenance	2–3	Routine maintenance
Waterproof membrane	Punching. Wear, tear.	3	Timely repairs Replacement

1.3.4.6. *Waterproofing and fireproofing*

Type of defect	Probables causes	Severity index	Repair solution
Degradation of waterproof coatings	Aggressive environment UV	3–3P	Repair or replacement
Degradation of fire-resistant coatings	Aggressive environment UV	3–3P	Repair or replacement

1.4. Maritime structures

1.4.1. *General information*

As the approach used for inspections such as it has been indicated above for civil engineering structures is difficult to apply to port and maritime structures, the CETMEF has proposed a simplified method called the comparative simplified visit (CSV) method.

The principle of the method described in the CETMEF guide is based on the following actions:

– ascertain a nomenclature of port heritage as has been done for other types of civil engineering structures;

– establish an inspection plan including visits to define the mechanical state and the state of use of the various structures listed in the nomenclature;

– prioritize levels of degradation and vulnerability and thus establish a plan of priorities;

– plan the diagnostics required and any necessary reinforcement work.

This guide may apply in particular to:

– docks (on piles, in caisson, etc.);

– pontoons and moorings;

– dikes;

– riprap;

– footbridges, locks, etc.

The specificities of the maritime environment lie mainly in the aggressiveness of the environment with respect to concrete and steel:

– physicochemical aggressiveness of seawater that contains both chlorides and sulfates and is more or less sensitive depending on the exposure area (submerged zone, low water zone, splash zone, tide and spray zone). Eurocode 3 defines the values presented in Table 1.10;

– mechanical aggressiveness, particularly due to swell.

Duration of use of the project	5 years	25 years	50 years	75 years	100 years
Ordinary fresh water (river, navigable canal, etc.) in the high attack zone (water line)	0.15	0.55	0.90	1.15	1.40
Heavily polluted freshwater (wastewater, industrial effluents, etc.) in the high attack zone (water line)	0.30	1.30	2.30	3.30	4.30
Sea water under temperate climate in the high attack zone (low water and spray zone)	0.55	1.90	3.75	5.60	7.50
Seawater in temperate climates in the permanent immersion zone or in the tidal zone	0.25	0.90	1.75	2.60	3.50
NOTES.– 1) The highest corrosion rate is usually found in the spray zone or in the low water area. However, in most cases, the highest bending moment is in the permanent immersion zone. 2) The values given for 5 and 25 years are based on measurements, while the other values are extrapolated.					

Table 1.10. *Corrosion sacrificial thickness according to EC3*

The forces are defined in the "Recommendations for calculation at the limit states of maritime structures" (so-called Rosa 2000 recommendations).

– mechanical aggressiveness of mooring and docking of boats;

– mechanical aggressiveness of port equipment (cranes, etc.);
– chemical aggressiveness of products stored on the platforms.

1.4.2. *Principles of the CSV method*

The CSV method can be summarized in Figure 1.7.

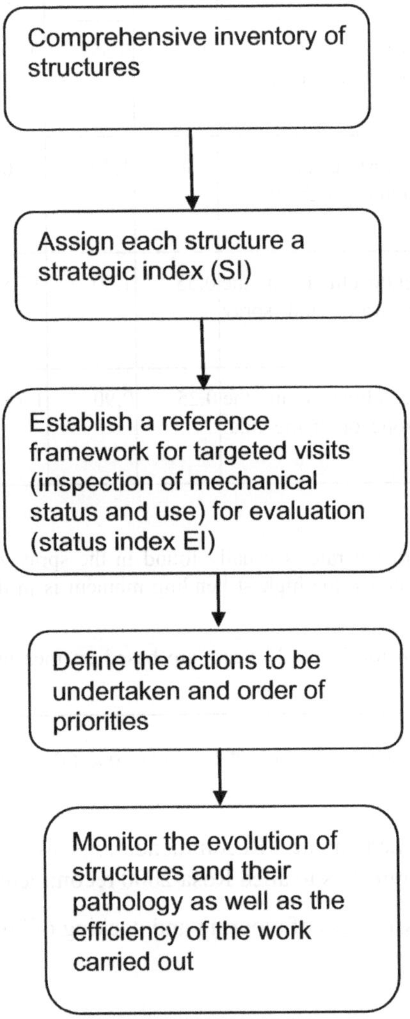

Figure 1.7. *Principle of the CSV method*

During the inspection of the structures, each element is allocated:

– *a mechanical state index (EIm)*, which can vary from 1 to 4 as presented in Table 1.11.

EIm index	Evaluation of the state
1	Structures with severe mechanical damage with risk of immediate ruin
2	Structures with serious mechanical damage without risk of immediate ruin
3	Structures with minor degradation or pathology
4	Structure in good condition

Table 1.11. *Evaluation of the mechanical state table*

– *a status indicator (EIu)* to evaluate operating conditions and safety problems relating to use. During the inspection, each structure is graded from 1 to 4 (Table 1.12)

EIu index	Evaluation of the state
1	Elements of use presenting degradations capable of generating immediate safety problems
2	Elements of use presenting degradations likely to generate operating problems
3	Elements of use presenting degradations likely to generate discomfort problems
4	Elements of use in good condition

Table 1.12. *Evaluation of the state table*

The status index of the structures is then defined as:

$EI = Min\ (EIm;\ EIu)$

The actions to be carried out are then defined according to the EI value (Table 1.13).

EI index	Actions to be taken
1	– Prohibition of access and operation – Information on the risk of ruin – Temporary safety works (purging, etc.) – Complete diagnosis of the structure – Monitoring
2	– Additional diagnostics including detailed inspection, material testing, underwater inspections, etc. – Monitoring of the structure – Study of the structural reinforcement project – Implementation of reinforcement or demolition works
3	– Additional diagnostics including detailed inspection, material testing, underwater inspections, etc. – Monitoring of the structure – Study of the structural repair project – Repair and specialized maintenance of works (painting, etc.)
4	Maintenance of the structure in good condition through: – Cleaning – Routine maintenance

Table 1.13. *Actions to be taken table*

1.4.3. *Determination of the strategic index SI*

The SI index is defined as "the value of strategic importance of the structure within the heritage".

Strategic decisions can be made by:

– a group of structures (set of structures with the same general use); a classification is then established by the manager (for example swell protection structures that are more strategic than the wharves);

– a family of structures (for example a family of unloading stations may be more strategic than a family of wharves);

– the structures directly.

In the CSV method, each structure is assigned:

– a name;

– a location;

– a specific function.

The decision-making criteria defined in the CETMEF guide are generally the following:

– the value added of landed goods;

– passenger traffic on the structure;

– the possibility of by-passing the structure;

– the lost value added;

– the value of the new structure;

– the heritage value of the structure;

– the strategic nature;

– ease of repair.

For each criterion, the manager gives a score of 1–4 (for example value added: (1) significant, (2) average, (3) low, (4) very low).

1.4.4. *Frequency of visits*

The periodicities are generally the following:

– mechanical visits: between 3 and 5 years;

– usage visits: between 6 months and 1 year.

1.4.5. *Defining the priorities*

The definition of priorities is obtained by crossing the SI and EI indices.

This crossing is done by the manager of maritime structures within the context of a risk analysis.

An example is given in the CETMEF document in Figure 1.8.

Management emphasizing the mechanical aspect

Decreasing priority

	EI	SI
A	1	1
B	1	2
C	1	3
D	1	4
E	2	1
F	2	2
G	2	3
H	2	4
I	3	1
J	3	2
K	3	3
L	3	4
M	4	1
N	4	2
O	4	3
P	4	4

Management emphasizing the strategic aspect (*)

Decreasing priority

	EI	SI
A	1	1
B	1	2
C	1	3
D	1	4
E	2	1
I	3	1
M	4	1
F	2	2
J	3	2
N	4	2
I	2	3
L	3	3
O	4	3
J	2	4
M	3	4
P	4	4

(*) Note: Even with a management policy that favors strategy, structures with a Status Index equal to 1 have priority because they require immediate safety.

Figure 1.8. *Example of management (source: CETMEF). For a color version of this figure, see "http://www.iste.co.uk/Lauzin/engineering.zip*

1.4.6. *Summary of the CSV method*

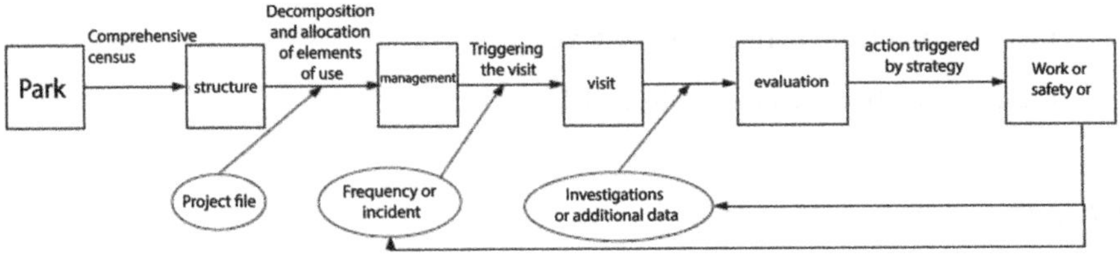

Figure 1.9.

NOTE.– An example of use can be found in Appendix 3.

1.4.7. *Points to look out for*

1.4.7.1. *Structure weight of masonry*

Type of defect	Probable causes	EIm severity index	Repair solution
Alteration of the laying mortar	Mechanical wave action Chemical action of sulfates in sea water	1–3	Replenishing joints Sprayed mortar (see Chapter 4, section 4.1.4)
Deformation of the cladding	Pressure on the wall Overload on the median Alteration of the laying mortar	1–3	Complete wall recovery Injection behind the curtain (see Chapter 4, section 4.3) Limitation of overloading
Vertical or oblique crack	Differential compaction Scouring	1–3	Underpinning (see Chapter 4, section 4.3)
Horizontal crack	Pressure on the wall Alteration of the laying mortar	1–3	Sprayed mortar (see Chapter 4, section 4.1.4) Limitation of overloading Injection behind the sheet (see Chapter 4, section 4.3)
Opening in the compartment parallel to the wall	Wall tilting Foundation scouring Large sliding circle	1–2	Underpinning (see Chapter 4, section 4.3) Installation of tie rods
Lack of support of superficial foundations	Scouring Too much dredging	1–2	Underpinning (see Chapter 4, section 4.3)
Wall tilting	Load too high Foundation scouring Excessive soil stress Shocks or moorings	1–2	Underpinning (see Chapter 4, section 4.3) Installation of tie rods Limitation of overloading Prohibition of mooring
Sliding of the base of the structure	Earth pressure Overload on the median Subdimensioning of the foundation	1–2	Underpinning (see Chapter 4, section 4.3) Installation of tie rods Limitation of overloading

1.4.7.2. *Concrete weight structure*

Type of defect	Probable causes	EIm severity index	Repair solution
Alteration of concrete	Mechanical wave action Chemical action of sulfates in sea water ISR Accidental actions (mooring, etc.) See Chapter 3, section 3.1	1–3	Replacement of the wall Shotcrete (see Chapter 4, section 4.1.4)
Deformation of the cladding	Pressure on the wall Overloading on the median	1–3	Complete wall recovery Injection behind the curtain (see Chapter 4, section 4.3) Limitation of overloading
Vertical or oblique crack	Differential compaction Scouring Restraint of concrete	1–3	Underpinning (see Chapter 4, section 4.3)
Horizontal crack	Pressure on the wall Scourings	1–3	Underpinning (see Chapter 4, section 4.3) Limitation of overloading Injection behind the sheet (see Chapter 4, section 4.3)
Opening in the compartment parallel to the wall	Wall tilting Scouring of the foundations Large sliding circle	1–2	Underpinning (see Chapter 4, section 4.3) Installation of tie rods
Lack of support of superficial foundations	Scouring Too much dredging	1–2	Underpinning (see Chapter 4, section 4.3)
Wall tilting	Excessive load restrained Scouring of the foundations Excessive soil stress Shocks or moorings	1–2	Underpinning (see Chapter 4, section 4.3) Installation of tie rods Limitation of overloading Prohibition of mooring
Sliding of the base of the structure	Earth pressure Overloading on the median Undersizing of the foundation	1–2	Underpinning (see Chapter 4, section 4.3) Installation of tie rods Limitation of overloading

1.4.7.3. *L-shaped reinforced concrete wall*

Type of defect	Probable causes	Elm severity index	Repair solution
Alteration of concrete	Mechanical wave action Chemical action of sulfates in sea water ISR Accidental actions (mooring, etc.) See Chapter 3, section 3.1	1–3	Replacement of the wall Shotcrete (see Chapter 4, section 4.1.4)
Visible steel Corrosion	Coating defect Shock Chloride attack (see Chapter 3, section 3.1)	2–3	Timely repair (see Chapter 4, section 4.1.5) Shotcrete (see Chapter 4, section 4.1.4) Cathodic protection (see Chapter 4, section 4.2)
Deformation of the cladding	Pressure on the wall Overloading on the median	1–3	Complete recovery of the wall Injection behind the sheet (see Chapter 4, section 4.3) Limitation of overloading
Vertical or oblique crack	Differential settlement Scourings Restraint of concrete	1–3	Underpinning (see Chapter 4, section 4.3)
Horizontal crack	Pressure on the wall Scourings	1–3	Underpinning (see Chapter 4, section 4.3) Limitation of overloading Injection behind the sheet (see Chapter 4, section 4.3)
Opening in the compartment parallel to the wall	Wall tilting Scouring of the foundations Large sliding circle	1–2	Underpinning (see Chapter 4, section 4.3) Installation of tie rods
Lack of support of superficial foundations	Scouring Too much dredging	1–2	Underpinning (see Chapter 4, section 4.3)
Wall tilting	Excessive load restrained Scouring of the foundations Excessive soil stress Shocks or moorings	1–2	Underpinning (see Chapter 4, section 4.3) Installation of tie rods Limitation of overloading Prohibition of mooring
Sliding of the base of the structure	Earth pressure Overloading on the median Under-sizing of the foundation	1–2	Underpinning (see Chapter 4, section 4.3) Installation of tie rods Limitation of overloading

1.4.7.4. *Sheet piling*

Type of defect	Probable causes	EIm severity index	Repair solution
Corrosion of the sheet	Corrosion protection defect Water pollution	2–3	Cathodic protection (see Chapter 4, section 4.2) Replacement Create a new sheet in front of the old one
Rips in the sheet Unpicking of locks	Mechanical actions Stresses greater than those calculated Failure to comply (threshing)	1–3	Limit overloading Welding of the keys
Deformation in the sheet plane	Stresses greater than those in the calculations Excessive stretching or breaking of tie rods Anchorage length is too weak Drainage defect behind the sheet	2–3	Limit operating loads Replacement of tie rods or implementation of a new bed of tie rods Increase foot stop Provide drainage
Deformation in the plane perpendicular to the sheet	Lack of ground bearing capacity Vertical stresses greater than those in the calculations	1–2	Underpinning (see Chapter 4, section 4.3) Limit overloading on the sheet
Tilting of the sheet toward the ground	Sliding of the bottom of the wall Stresses greater than those in the calculations Failure to comply	1–2	Restore the foot stop Limitation of overloading Prohibition of mooring
Tilting of the sheet toward the sea	Anchorage length is too weak Failure to comply Detension or rupture of tie rods Scouring or excessive dredging Stresses greater than those in the calculations	1–2	Underpinning (see Chapter 4, section 4.3) Installation of tie rods Limitation of overloading Prohibition of mooring Restore the foot stop

Land collapse behind the sheet	Fox phenomenon (bringing about ores) Burst pipe	2–3	Injection behind the sheet (see Chapter 4, section 4.3) Recovery of pipes
Land compaction behind the sheet	Natural consolidation Fox phenomenon Burst pipe	2–3	Injection behind the sheet (see Chapter 4, section 4.3) Recovery of pipes
Ground cracking behind the sheet	Sheet deformation Large sliding circle	1–3	Anchor rods
Alteration of the piercap	Mechanical wave action Chemical action of sulfates in sea water ISR Accidental actions (mooring, etc.) See Chapter 3, section 3.1	1–3	Replacement of the wall Shotcrete (see Chapter 4, section 4.1.4)
Visible steel Corrosion	Coating defect Shock Chloride attack (see Chapter 3, section 3.1)	2–3	Timely repair (see Chapter 4, section 4.1.5) Shotcrete (see Chapter 4, section 4.1.4) Cathodic protection (see Chapter 4, section 4.2)

1.4.7.5. *Diaphragm walls*

Type of defect	Probable causes	EIm severity index	Repair solution
Concrete degradation	Mechanical wave action Chemical action of sulfates in sea water ISR Accidental actions (mooring, etc.) See Chapter 3, section 3.1	1–3	Replacement of the wall Shotcrete (see Chapter 4, section 4.1.4)

Degradation of panel joints	Mechanical actions Mechanical wave action Chemical action of sulfates in sea water ISR Accidental actions (mooring, etc.) See Chapter 3, section 3.1 Failure to comply	2–3	Injection (see Chapter 4, section 4.3) Prohibit mooring
Cracking in the piercap	Thermal actions Shrinkage Stresses greater than those in the calculations See Chapter 3, section 3.1	1–3	Injection of cracks (see Chapter 4, section 4.1.5) Shotcrete (see Chapter 4, section 4.1.4) Limit operational overloading
Deformation in the curtain plane	Stresses greater than those in the calculations Excessive stretching or breaking of tie rods Anchorage length is too weak Drainage defect behind the sheet	2–3	Limit operating loads Replacement of tie rods or implementation of a new bed of tie rods Increase foot stop Provide drainage
Deformation in the plane perpendicular to the curtain	Lack of ground bearing capacity Vertical stresses greater than those in the calculations	1–2	Underpinning (see Chapter 4, section 4.3) Limit overloading on the sheet
Tilting of the curtain toward the ground	Sliding of the bottom of the wall Stresses greater than those in the calculations Failure to comply	1–2	Restore the foot stop Limitation of overloading Prohibition of mooring
Tilting of the ~~sheet~~ curtain toward the sea	Anchorage length is too weak Failure to comply Detension or rupture of tie rods Scouring or excessive dredging Stresses greater than those in the calculations	1–2	Underpinning (see Chapter 4, section 4.3) Installation of tie rods Limitation of overloading Prohibition of mooring Restore the foot stop
Land collapse behind the curtain	Fox phenomenon (bringing about ores) Burst pipe	2–3	Injection behind the sheet (see Chapter 4, section 4.3) Recovery of pipes
Land compaction behind the curtain	Natural consolidation Fox phenomenon (runoff of fine soil particles) Burst pipe	2–3	Injection behind the sheet (see Chapter 4, section 4.3) Recovery of pipes
Ground cracking behind the curtain	Sheet deformation Large sliding circle	1–3	Anchor rods

1.4.7.6. *Concrete dock on piles (metal or concrete)*

Type of defect	Probable causes	EIm severity index	Repair solution
Concrete degradation	Mechanical wave action Chemical action of sulfates in sea water ISR Accidental actions (mooring, etc.) See Chapter 3, section 3.1	1–3	Replacement of the wall Shotcrete (see Chapter 4, section 4.1.4)
Alteration of reinforcements Corrosion of metal structures	Chemical action of chlorides Cathodic protection defect See Chapter 3, section 3.1	2–3	Timely repair (see Chapter 4, section 4.1.5) Shotcrete (see Chapter 4, section 4.1.4) Cathodic protection (see Chapter 4, section 4.2)
Cracking of the platform slab	Thermal actions Shrinkage Stresses greater than those in the calculations Defect in bearing capacity of piles See Chapter 3, section 3.1	1–3	Injection of cracks (see Chapter 4, section 4.1.5) Shotcrete (see Chapter 4, section 4.1.4) Limit operational overloading Underpinning (see Chapter 4, section 4.3)
Settlement of piles	Lack of ground bearing capacity Vertical stresses greater than those in the calculations	1–2	Limit operational overloading Underpinning (see Chapter 4, section 4.3)
Cracking of the ground behind the platform	Sheet deformation Large sliding circle Compaction	1–3	Anchor rods Injection behind the sheet (see Chapter 4, section 4.3)

1.4.7.7. *Rockfill wharf*

Type of defect	Probable causes	Elm severity index	Repair solution
Concrete degradation	Mechanical wave action Chemical action of sulfates in sea water ISR See Chapter 3, section 3.1	1–3	Replacement of the wall Shotcrete (see Chapter 4, section 4.1.4)
Sag in the coating	Cavities Leakage of fine materials Compaction	1–3	Injection behind the sheet (see Chapter 4, section 4.3) Replacement Recovery in the underground (see Chapter 4, section 4.3)
Vertical or oblique crack in the cladding	Compaction Cavities	1–3	Injection behind the sheet (see Chapter 4, section 4.3) Replacement Underpinning (see Chapter 4, section 4.3)
Horizontal crack in the cladding	Stresses in the cladding greater than those in the calculation Compaction	1–3	Sprayed mortar (see Chapter 4, section 4.1.4) Limitation of overloading Injection behind the sheet (see Chapter 4, section 4.3)
Cracking of the ground behind the riprap	Insufficient foot stop Large sliding circle Scouring	1–2	Underpinning (see Chapter 4, section 4.3) Installation of tie rods Restore the foot stop
Alteration of riprap	Mechanical wave action Chemical actions of seawater See Chapter 3, section 3.1	1–3	Replacement of the wall Sprayed mortar (see Chapter 4, section 4.1.4)

1.4.7.8. *Coffered reinforced concrete piers*

Type of defect	Probable causes	EIm severity index	Repair solution
Concrete degradation	Mechanical wave action Chemical action of sulfates in sea water ISR Accidental actions (mooring, etc.) See Chapter 3, section 3.1	1–3	Replacement of the wall Shotcrete (see Chapter 4, section 4.1.4)
Alteration of reinforcements Corrosion of metal structures	Chemical action of chlorides Cathodic protection defect See Chapter 3, section 3.1	1–3	Replacement of the wall Shotcrete (see Chapter 4, section 4.1.4) Resumption of cathodic protection (see Chapter 4, section 4.2)
Vertical or oblique cracking of the caissons	Thermal actions Shrinkage Stresses greater than those in the calculations Differential settlement See Chapter 3, section 3.1	1–3	Underpinning (see Chapter 4, section 4.3) Strengthening of structure (see Chapter 4, section 4.1.4) Limitation of overloading
Horizontal cracking of caissons or at block joints	Stresses greater than those in the calculations Scouring See Chapter 3, section 3.1	1–3	Underpinning (see Chapter 4, section 4.3) Limitation of overloading
Cracking of the ground behind the platform	Deformation of the wharf wall Large sliding circle Compaction	1–3	Underpinning (see Chapter 4, section 4.3) Limitation of overloading Injection behind the sheet (see Chapter 4, section 4.3)
Tilting	Scouring Excessive dredging	1–2	Underpinning (see Chapter 4, section 4.3) Installation of tie rods Limitation of overloading Prohibition of mooring
Sliding	Stresses greater than those in the calculations Foot stop defect	1–2	Underpinning (see Chapter 4, section 4.3) Installation of tie rods Limitation of overloading

Figure 1.10. *For a color version of this figure, see http://www.iste.co.uk/Lauzin/engineering.zip*

1.5. *Silos*

1.5.1. *General information*

Regarding silos (cereal storage facilities), INERIS has published an inspection and maintenance guide.

This guide mainly focuses on:

– storage of grain, flour;

– movable storage walls;

– fertilizer boxes;

– metal tanks;

– polyester tanks;

– retention basins (storage of liquid fertilizers, plant protection products, extinguishing water, etc.);

– various facilities (mill, reception pits, gallery, etc.);

– safety accessories.

In addition, the "Guide to Art on Silos" recalls that structures must be *monitored*.

1.5.2. *Reminder on the regulations for the mechanical operation of silos*

The loads that should be applied to the silo walls are described in EN 1991-4.

These loads mainly consist of:

– ensiled material;

– the weight of civil works.

The forces generated by the ensiled material and taken up by the walls of the silo depend on the following parameters (section 4.3 of EN 1991-4):

– specific weight of bulk material γ;

– wall friction coefficient μ;

– internal friction angle φi;

– coefficient of lateral constraint K;

– cohesion C;

– coefficient of localized pressure C_{op}.

The forces are then summarized in Figure 1.11 (for slender silos).

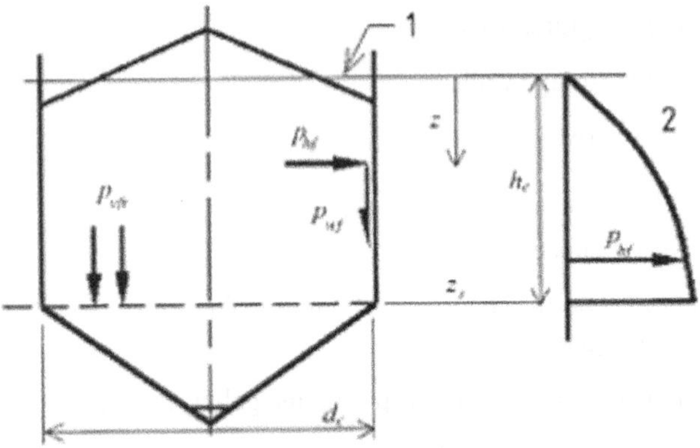

Legend

1 Surface area
2 Horizontal stress along the vertical wall section

Figure 1.11. *Silo forces*

These forces depend on the parameters above, so any change in the initial hypotheses (change in silage material, modification of the friction coefficient on the walls, etc.) must be justified.

These various parameters are measured for each test and are summarized in Table 1.14.

1.5.3. *Principle of inspection*

Similar to the aforementioned structures, the inspection plan can be divided into several stages in the following manner:

– an inspection (level 1) to ascertain the state of the structure from visual observations, which will or will not trigger a level 2 visit;

– a more targeted inspection of the critical pathological points (level 2), which will establish the causes of the damage and possible remedial solutions;

– a level 3 inspection with experts on this type of structure if the previous two visits did not reach a formal conclusion.

Type of material d), e)	Specific weight b) γ (in kN/m³)		Angle of rest Φ_r (in degrees)	Internal friction angle Φ_i (in degrees)		Lateral coefficient of stress K		Coefficient of wall friction c) μ ($\mu = \tan \varphi_w$)				Reference coefficient for localized load
	γ_l	γ_u		φ_{im}	a_φ	K_m	a_K	type D1	type D2	type D3	a_μ	C_{op}
	inferior	additional		average	factor	average	factor	average	average	average	factor	
Default material a)	6,0	22,0	40	35	1,3	0,50	1,5	0,32	0,39	0,50	1,40	1,0
Aggregates	17,0	18,0	36	31	1,16	0,52	1,15	0,39	0,49	0,59	1,12	0,4
Alumina	10,0	12,0	36	30	1,22	0,54	1,20	0,41	0,46	0,51	1,07	0,5
Mixtures for animal feed	5,0	6,0	39	36	1,08	0,45	1,10	0,22	0,30	0,43	1,28	1,0
Tourteau	6,5	8,0	37	35	1,06	0,47	1,07	0,23	0,28	0,37	1,20	0,7
Barley ○	7,0	8,0	31	28	1,14	0,59	1,11	0,24	0,33	0,48	1,16	0,5
Cement	13,0	16,0	36	30	1,22	0,54	1,20	0,41	0,46	0,51	1,07	0,5
Clinker of cement b)	15,0	18,0	47	40	1,20	0,38	1,31	0,46	0,56	0,62	1,07	0,7
Coal ○	7,0	10,0	36	31	1,16	0,52	1,15	0,44	0,49	0,59	1,12	0,6
Coal dust ○	6,0	8,0	34	27	1,26	0,58	1,20	0,41	0,51	0,56	1,07	0,5
Coke	6,5	8,0	36	31	1,16	0,52	1,15	0,49	0,54	0,59	1,12	0,6
Fly ash	8,0	15,0	41	35	1,16	0,46	1,20	0,51	0,62	0,72	1,07	0,5
Flour ○	6,5	7,0	45	42	1,06	0,36	1,11	0,24	0,33	0,48	1,16	0,6
Iron mineral pellets	19,0	22,0	36	31	1,16	0,52	1,15	0,49	0,54	0,59	1,12	0,5
Hydrated lime	6,0	8,0	34	27	1,26	0,58	1,20	0,36	0,41	0,51	1,07	0,6
Limestone powder	11,0	13,0	36	30	1,22	0,54	1,20	0,41	0,51	0,56	1,07	0,5
Corn ○	7,0	8,0	35	31	1,14	0,53	1,14	0,22	0,36	0,53	1,24	0,9
Phosphate	16,0	22,0	34	29	1,18	0,56	1,15	0,39	0,49	0,54	1,12	0,5
Potatoes	6,0	8,0	34	30	1,12	0,54	1,11	0,33	0,38	0,48	1,16	0,5
Sand	14,0	16,0	39	36	1,09	0,45	1,11	0,38	0,48	0,57	1,16	0,4
Clinker of slag	10,5	12,0	39	36	1,09	0,45	1,11	0,48	0,57	0,67	1,16	0,6
Soybeans	7,0	8,0	29	25	1,16	0,63	1,11	0,24	0,38	0,48	1,16	0,5
Sugar ○	8,0	9,5	38	32	1,19	0,50	1,20	0,46	0,51	0,56	1,07	0,4
Beet pellets	6,5	7,0	36	31	1,16	0,52	1,15	0,35	0,44	0,54	1,12	0,5

Table 1.14. *EN 1991-4: Appendix C*

Following the level 2 inspection, each structure or part of the structure is assigned an index of damage (Table 1.15).

Level	Class of defects	Description of the level	Follow-up (type of investigation)
1	d1	Structures in good condition for which any damage can be repaired through conventional maintenance	Nothing in particular to report, follow-up and normal maintenance of the structure (annual) Periodic inspection
2	d2	Defects that can be repaired through specialized maintenance or that can evolve over time	Repairs to consider. Maintenance plan to be updated
3	d3	Damage that may call into question the general or local stability of the structure. A priority building given the amount of damage.	Immediate repair. Put under surveillance. Security perimeter.

Table 1.15. *Level of inspection table*

This hierarchy of damages then makes it possible to classify the structure (Table 1.16).

Class of structure	Description of the level
1	Structure without any level d2 or d3 damage.
2	A structure without any level d3 damage but capable of presenting level d2 damage.
3	Structure with level d3 damage

Table 1.16. *Class of structure table*

For the inspection, we refer the reader to sections 2.3 and 2.4.

1.5.4. *Follow-up file*

The purpose of the follow-up file is to provide a good knowledge of a structure from its construction with the history of any interventions that have been carried out (maintenance, equipment works, structural modifications, etc.).

This file should include at least:

– the implementation plans (formwork, reinforcement, materials, etc.), everything that constitutes the aforementioned project file for the structures;

– the activity of the installation during its design (cereal silo, transfer silo, etc.);

– the current activity of the structure (if modified in relation to the initial activity);

– the characteristics of the use (rate of rotation, etc.);

– inspection sheets that have already been completed (level 1 and 2);

– modifications or repairs that have been undertaken (reinforcement, opening of chute, etc.);

– the protective coatings used (inner resin, exterior coating, cathodic protection, etc.);

– incidents that have occurred;

– safety equipment (footbridges, guardrails, etc.).

1.5.5. *Inspection procedure*

The role of the inspector in carrying out the inspection program is to:

– establish access means and equipment;

– survey the meteorological conditions (rain, wind, snow, ice, etc.) with indications of the temperatures;

– run a statement of special conditions;

– conduct a close visual inspection detailing the defects encountered. Any defect shall be characterized by:

- its type (crack, spalling, etc.);
- its physical and dimensional characteristics;

- its scope;
- its location.

The observations to be made on-site and to be recorded in the inspection report are:

– the zone of influence (buried environment area, aerial zone, etc.);

– the silo (area in contact with the ensiled material, area in contact with the air, area in contact with the ground);

– equipment (guard rails, ladders, cover seal, optional silo interior seal, etc.);

– cell support posts and sails;

– the foundations;

– approach and access;

– the features of the structure.

1.5.6. *The inspection report*

The inspection report must include:

– a chapter identifying the structure;

– a chapter specifying the general characteristics;

– a chapter containing information on the design and operation of the structure;

– a chapter on the life of the structure;

– a chapter on the findings and measurements carried out as part of the inspection;

– a chapter on tests, auscultations, investigations;

– a summary chapter on the state of the structure and its evolution;

– appendices with:

- the plans of the structure (longitudinal, transverse, elevation);
- plans and patterns of pathologies encountered;
- photographic report.

1.5.7. *Points to look out for*

Type of defect	Probable causes	Severity index	Repair solution
Visible steel	Cover defect (see Chapter 3, section 3.1)	Localized d1 Widespread d2d/3	Timely repair (see Chapter 4, section 4.1.5) Shotcrete (see Chapter 4, section 4.1.4) Cathodic protection (see Chapter 4, section 4.2)
Segregation	Formwork sealing failure. Poor implementation of concrete. Inadequate rheology.	d1	Timely repair (see Chapter 4, section 4.1.5)
Disintegration of concrete	Quality of concrete not adapted to the environment Implementation defect (see Chapter 3, section 3.1)	d2/d3	Depending on the results of the chemical analysis of the concrete (see Chapter 4)
Concrete peeling	Shock Aggressive environment Quality of concrete (see Chapter 3, section 3.1)	Non-changing d1 Changing d2	Timely repair (see Chapter 4, section 4.1.5) Shotcrete (see Chapter 4, section 4.1.4)
Faience	Shrinkage Alkali-reaction Internal sulfate reaction (see Chapter 3, section 3.1)	d1 d2/d3 d2/d3	Shrinkage: protection by technical paint (I3, I4) For an alkali-reaction or ISR: to be seen depending on chemical analyses
Isolated cracks w ≤ 2/10 mm 2/10< w ≤ 20/10 20/10< w	See Chapter 3, section 3.1 Check whether the crack has changed	D1 d2/d3 d3p	Timely repair (see Chapter 4, section 4.1.5) in the first two cases To analyze in the third case (lizards)
Multiple cracks w ≤ 2/10 mm 2/10< w ≤ 20/10 20/10< w	See Chapter 3, section 3.1 Check whether the crack has changed.	D1 d2/d3 d3p	Timely repair (see Chapter 4, section 4.1.5) Shotcrete (see Chapter 4, section 4.1.4) Additional prestressing (see Chapter 4, section 4.1.3)
Cracks under loading or unloading	Sizing failure (dynamic effects) (See Chapter 3, section 3.1, check whether the crack has changed)	d3	Shotcrete (see Chapter 4, section 4.1.4) Additional prestressing (see Chapter 4, section 4.1.3)

Visible foundations	Scouring, compaction of the soil around the silo	Localized d1 Widespread d2d/3	Backfill Underpinning (see Chapter 4, section 4.3)
Verticality defect	Differential compaction Hydrology of the site Compaction of the backfill Evacuation of stormwater	Stabilized without cracking d1 Stabilized with cracking d2 Non-stabilized d2 /d3	Underpinning (see Chapter 4, section 4.3)
Cracking of the udders (roundheads) Visible steel Petal roundheads Opening between elements	Cover defect (see Chapter 3, section 3.1) Pipe passage Failure to comply	Localized d1 Widespread d2d/3 d2/d3	Timely repair (see Chapter 4, section 4.1.5) Shotcrete (see Chapter 4, section 4.1.4) Cathodic protection (see Chapter 4, section 4.2) Strapping Shotcrete (see Chapter 4, section 4.1.4)

EXAMPLES.–

Figure 1.12. *Opening of the skirt of the silo following the implementation of an internal lining*

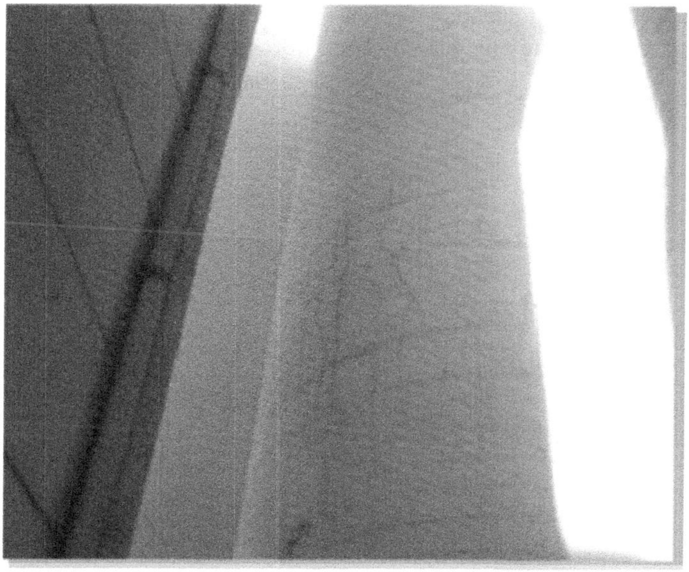

Figure 1.13. *Vertical cracking of the skirt of the cylindrical silo*

1.6. Gantry, metal hanger and high masts

1.6.1. *General information*

The topic discussed here is mainly that of concrete and metallic structures as defined in the LCPC technical guide "Gantry, metal hanger, high masts".

1.6.2. *Principle of inspection*

Similarly to the aforementioned structures, the inspection plan can be divided into several stages in the following manner:

– an inspection (level 1) to ascertain the state of the structure from visual observations, which will or will not trigger a level 2 visit;

– a more targeted inspection of the critical pathological points (level 2), which will establish the causes of the damage and the possible remedial solutions;

– a level 3 inspection with experts on this type of structure if the previous two visits did not reach a formal conclusion.

The guide does not mention the classification of structures; however, it is possible to approximate the classification for the structures, as mentioned in section 1.7.

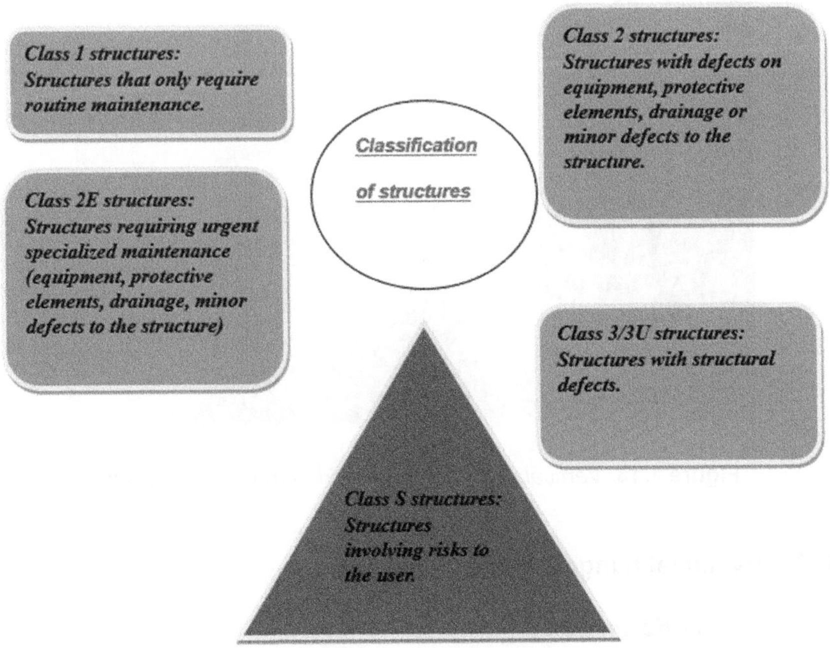

Figure 1.14. *Classification of structures*

1.6.3. *The inspection report*

The inspection report must include:

– a chapter identifying the structure;

– a chapter specifying the general characteristics;

– a chapter containing information on the design and operation of the structure;

– a chapter on the life of the structure;

– a chapter on the findings and measurements carried out as part of the inspection;

– a chapter on tests, auscultations, investigations;

– a summary chapter on the state of the structure and its evolution;

– Appendices with:

- the plans of the structure (longitudinal, transverse, elevation);

- plans and patterns of pathologies encountered;

- photographic report.

NOTE.– An example of a report can be found in Appendix 4.

1.6.4. *Points to look out for*

Type of defect	Probable causes	Severity index	Repair solution
Mass			
Visible steel	Cover defect (see Chapter 3, section 3.1)		Timely repairs (see Chapter 4, section 4.1.5) Shotcrete (see Chapter 4, section 4.1.4) Cathodic protection (see Chapter 4, section 4.2)
Segregation	Formwork sealing defect Poor implementation of concrete Inadequate rheology		Timely repairs (see Chapter 4, section 4.1.5)
Disintegration of concrete	Quality of concrete not adapted to the environment Implementation defect (see Chapter 3, section 3.1)		Depending on the results of the chemical analysis of the concrete (see Chapter 4)

Concrete peeling	Shock Aggressive environment Quality of concrete (see Chapter 3, section 3.1)		Timely repairs (see Chapter 4, section 4.1.5) Shotcrete (see Chapter 4, section 4.1.4)
Faience	Shrinkage Alkali-reaction Internal sulphate reaction (see Chapter 3, section 3.1)		Shrinkage: protection through technical paint (I3, I4) In the case of alkali-reaction or ISR to be seen depending on chemical analyzes.
Isolated cracks ($w \leq 2/10$ mm $2/10 < w \leq 20/10$ $20/10 < w$)	Check for any change in the condition of the crack (see Chapter 3, section 3.1)		Timely repairs (see Chapter 4, section 4.1.5) in the first 2 cases. To analyze in the 3rd case (lizards)
Multiple cracks ($w \leq 2/10$ mm $2/10 < w \leq 20/10$ $20/10 < w$)	Check for any change in the condition of the crack (see Chapter 3, section 3.1)		Timely repairs (see Chapter 4, section 4.1.5) Shotcrete (see Chapter 4, section 4.1.4) Additional prestressing (see Chapter 4, section 4.1.3)

<table>
<tr><td>Cracks under loading or unloading</td><td>Sizing failure (dynamic effects)
(see Chapter 3, section 3.1)

Check for any change in the condition of the crack</td><td></td><td>Shotcrete
(see Chapter 4, section 4.1.4)

Additional prestressing
(see Chapter 4, section 4.1.3)</td></tr>
<tr><td>Visible foundation</td><td>Scouring, soil compaction</td><td></td><td>Backfill

Underpinning (see Chapter 4, section 4.3)</td></tr>
<tr><td>Verticality defect</td><td>Differential settlement

Hydrology of the site

Compaction of the embankment
Evacuation of storm water</td><td></td><td>Underpinning (see Chapter 4, section 4.3)</td></tr>
<tr><td colspan="4">Baseplate</td></tr>
<tr><td colspan="2">Condition of the baseplate</td><td colspan="2">Presence of eath dirt
Water retention
Degraded anti-corrosion protection
Corrosion
Deformation of the plate
Deformation of the gussets</td></tr>
<tr><td colspan="2">Welding state</td><td colspan="2">Cracks
Deblocking blow-holes
Lack of material
Corrosion
Evolutionary defect</td></tr>
</table>

Column	
General appearance	Geometric defect Localized deformation Shock Degraded corrosion protection Corrosion
Verification of welds	Cracks Deblocking blow-holes Lack of material Corrosion
Access hatch	Presence of hatch Presence of closing elements Watertightness of the hatch Condition of the bolt
Transom beam of frame	
General appearance	Geometric defect Localized deformation Shock Degraded corrosion protection Corrosion Link between column and beam
Verification of welds	Cracks Deblocking blow-holes Lack of material Corrosion
Access hatch	Presence of hatch Presence of closing elements Watertightness of the hatch Condition of the bolt

Traffic signs	
General appearance	Geometric defect Localized deformation Shock Degraded corrosion protection Corrosion
Verification of mechanical-fixings	Nature of the fixings Number of missing elements Number of loose elements Presence of locking nuts Corrosion
Access hatch	Presence of hatch Presence of closing elements Watertightness of the hatch Condition of the bolt

2

Concept of Resistance of Materials: Application to Reinforced Concrete

2.1. General information on reinforced concrete

The discovery of cement is often attributed to the Romans. The latter were undoubtedly the first to mix volcanic ashes from the Pozzuoli region with lime and see that the mixture thus formed hardened in the presence of water. They were able to use this mixture in masonry to bind stones together. It was at that time that the "mason's trowel" made an appearance.

Curiously, this discovery remained dormant through the Middle Ages and only reappeared in the 18th Century, in 1756 to be precise, in the works of the English engineer John Smeaton. He rediscovered the properties of clay in calcareous stones. An industrialization of this cement was then started by lime producers Parker and Wyats around 1786.

In France, it was not until 1817 that work by Louis Vicat brought to light a theory on the hydraulicity of lime and mortar.

In 1824, the English engineer Joseph Aspdin filed a patent for "Portland cement", the color of which was similar to that of the stone found in the quarries on the Portland peninsula.

Finally, in 1855, the French architect François Coignet built the first concrete building on rue Danton in Paris.

Known for its compressive strengths, the cement showed its weaknesses under other conditions of use. In 1845 came the idea of a cement-metal bonding, then called reinforced cement. The first example was the barque of Lambot, which was present at the universal exhibition of 1900.

The development of reinforced cement, then that of reinforced concrete, was important and saw its most significant example in the works of French engineer François Hennebique from 1879 onward. The latter was at the origin of structural inventions akin to the timber frame but made entirely of reinforced concrete.

From 1896 on, prefabricated housing projects made up of reinforced concrete began to make an appearance.

From the beginning of the 20th Century, traditional architecture was shattered by the discovery of this new material. On October 20, 1906, the first "ministerial instruction on the use of reinforced concrete" appeared. This regulation continued to evolve according to the new characteristics of the two main components of reinforced concrete: steel and cement. Therefore, when diagnosing an existing structure, it is essential to know the approximate age of the building and the regulations that were applicable at that time. In particular, verifications of the bearing capacity of reinforced concrete elements should be carried out in accordance with the regulations in force at the date of construction.

2.2. Concrete material

Concrete is a homogeneous mixture of the following components.

2.2.1. *Cement*

This consists of fine powders obtained by high-temperature firing and then grinding of a mixture of limestone and clay. With water, this mixture forms a paste that can "set" and gradually harden (hydraulic binder).

The choice of cement (type) and its dosage depend on the desired mechanical performance, on the resistance to possible aggressive agents and on the nature of other components.

Cements were defined by the revised French standard XP P 15-301 and then according to the European standard EN 197-1.

There are five broad categories of cement:

– Portland cement (CPA) that is based on clinker;

– composite Portland cement (CPJ) that is based on clinker with the addition of other components;

– blast furnace cement (CHF and CLK) that is based on slag (mineral residue from the preparation of cast iron in blast furnaces);

– pozzolanic cement (CPZ);

– slag and ash cement (CLC) that is based on clinker, fly ash and slag.

Besides these five main categories, there are other cement families that are not allowed to be used in reinforced concrete (hydraulic lime, XHN, etc.)

2.2.2. *Aggregates*

These are inert materials (sand, gravel, pebbles, etc.) that exist in the composition of concrete.

Generally, there are natural aggregates (rolled or crushed) and artificial aggregates (industrial or non-crushed industrial products such as crushed crystallized slag or granulated slag, etc.). Defined by EN 12620, aggregates are traditionally considered as the skeleton of concrete.

It is important to know their physicochemical and mechanical properties as well as their suitability for concrete.

2.2.3. *Mixing water*

The properties are given in EN 1008. This standard defines the physical and chemical properties of the mixing water.

2.2.4. *Admixture*

These are products used in small quantities that are capable of improving certain properties of concretes. For example, they can act on:

– the setting time;

– the mechanical properties;

– waterproofing;

– implementation, etc.

For existing structures, for example, we may note that from 1909 onward, sugar was used as a setting retarder. Then, between 1910 and 1920, we saw the appearance on the market of products based on calcium chloride (water repellent and setting accelerators).

In 1964, the COPLA (Permanent Commission for Liquid Binders and Concrete Admixers, from the French Commission Permanente des Liants hydrauliques et des Adjuvants du béton) was set up, which was in charge of establishing a list of products that could be used safely.

Later, in 1984, the NF adjuvants certification was introduced.

EN 934 classifies adjuvants into three main categories:

– those that modify the workability of concrete;

– those that act on the setting and hardening;

– those that modify certain mechanical properties.

NOTE.– *Fibers*: Only used in more recent times, fibers are used to reinforce the action of traditional reinforcements, in particular to oppose the propagation of microcracks. Glass fibers, metal fibers and polypropylene fibers are currently available on the market.

More recently, reactive powder concretes (RPCs) have been tested – the latest in Bouygues' scientific direction.

By analogy with the family of high-performance concrete, the aim of the study was to improve the homogeneity and capacity of the material.

2.2.5. *Mechanical properties of concrete*

In this chapter, we shall limit ourselves to discussing the definitive properties of concrete. Provisional characteristics such as maneuverability or segregation problems are primarily to do with implementation.

2.2.5.1. *Resistance*

Resistance is the most important property of concrete, and is an increasing function of the cement/water ratio and compactness.

By definition, concrete has good compressive strength but very low tensile strength. The compressive strength value is generally defined at 28 days and measured in destructive tests on broken cylindrical samples.

Because of the low tensile strength (a single crack can annul all resistance), it has been assumed since the first regulations (1906) that only the compressive strength of concrete is to be taken into account in the calculations.

For example, in the 1930s, gravel concrete normally dosed at 350 kg of Portland cement could have a compressive strength of 182 kg/cm² at 90 days (18 MPa).

In the 1960s, normally dosed concrete (350 kg/m³ CPA) could reach compressive strength values of 725 kg/cm² at 28 days (72.5 MPa), while RPC was tested at values ranging from 2,000 to 8,000 kg/cm² (200– 800 MPa).

2.2.5.2. *Shrinkage*

This is the phenomenon of reduction in size (general shortening) that accompanies the setting of concrete.

In his "New Guide to Concrete", Georges Dreux assimilates shrinkage to the effect of a lowering of temperature that results in shortening. Experimental studies have shown that hardening concrete in water greatly reduces the effects of shrinkage.

Therefore, it is common practice to water concrete parts (or use curing agents) during the hardening phase at a time when the concrete only has a low tensile strength and would crack easily under the effect of shrinkage.

Successive regulations have specified the conditions under which it is possible to ignore the effects of shrinkage (and variations in temperature) or the values to be taken into account in calculations.

For example, in 1932, Pugnet's experiments (published in the *Annales des Ponts et Chaussées*) showed tensile stresses of around 3–15 kg/cm² in concrete (0.3–1.5 MPa) depending on the percentage and storage conditions. These forces, which were added to those caused by permanent loads, the overloads, then had to be taken into account for measuring out the concrete elements.

Later, the BAEL 93 rules fixed construction lengths for which the effects of shrinkage and thermal expansion were not to be taken into account in calculations.

For example, in France, the shortening of concrete due to shrinkage is considered to be around:

$$\Delta l/l = 3 \times 10^{-4}$$

This shortening results in tensile stresses such as:

$$\Delta l/l = \sigma_b/E_b$$

where $\boldsymbol{\sigma_b}$ represents the tensile stress of the concrete only due to shrinkage and $\mathbf{E_b}$ represents the modulus of deformation of concrete (see below). It follows that:

$$\boldsymbol{\sigma_b = \Delta l/ l \cdot E_b} = 60 \text{ bar} = 6 \text{ MPa}$$

However, the tensile strength of concrete is much lower than 60 bar (20 bar). Therefore, concrete cracks, which legitimizes the assumption to not take account of tensile concrete in calculations.

2.2.5.3. *Creep*

This is the phenomenon of deferred deformation of concrete under constant load.

For the sake of comparison, old wooden floors can be used where the deformation of the main beams often reaches significant values.

Concrete has a plastic behavior over a certain load (about half the ultimate compressive strength). Even after removal of the load, the deformation remains. This deformation, which is due to the creep of concrete, continues over several months and can even continue over years.

2.2.5.4. *Thermal expansion*

We generally assume a coefficient of thermal expansion of 1×10^{-5}. This coefficient depends on the nature and qualities of the concrete, as well as the size of the aggregates. In France, a temperature variation of $\boldsymbol{\Delta\Theta} = \pm 20$ °C is commonly accepted, which implies a variation in length of:

$$\boldsymbol{\Delta l/l} = 2 \times 10^{-4}$$

2.2.5.5. *The deformation modulus E*

From the theory of elasticity where deformations are proportional to the applied stresses, the deformation modulus (or elasticity coefficient) is defined by:

E = Unit stress/relative strain

In light of the above (deformation due to creep), two deformation moduli are considered:

– an instant modulus;

– a deferred modulus;

to account for the fact that the total deformation (including creep effects) is about three times greater than instantaneous deformation.

This longitudinal deformation is accompanied by a transverse deformation (called the "Poisson effect"). The Poisson coefficient (ratio of transverse deformation to longitudinal deformation) has a value that is generally taken to be 0.2.

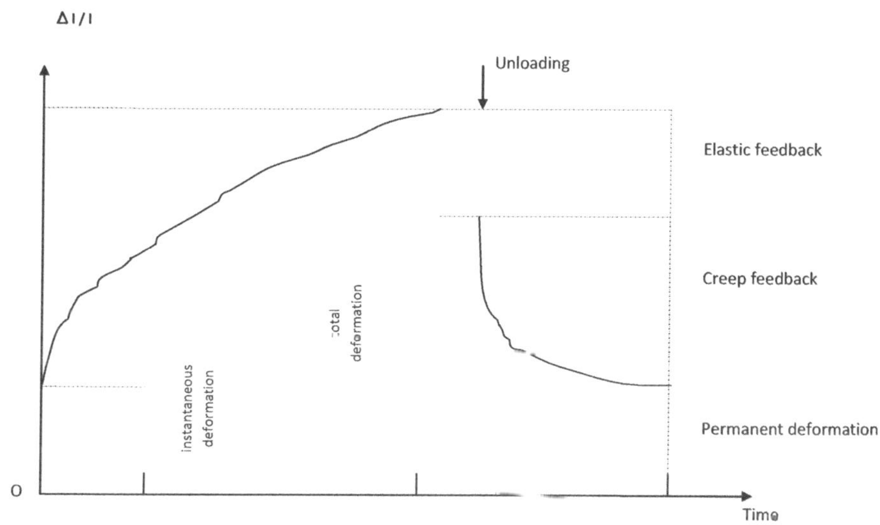

Figure 2.1. *Influence of creep on permanent deformation*

2.2.5.6. *The deformation – stress diagram*

This reflects the deformation mode of concrete as a function of the stress applied to the sample.

We have previously seen that the deformation modulus (or elasticity coefficient or Young's modulus) measured the capacity of concrete to deform under stress.

For example, in the 1935 regulations, it was considered that very meticulous concrete with a compressive strength of 250 kg/m² will break under a traction of 20 kg/cm².

2.2.6. *Eurocode 2 provisions for concrete*

Today, the provisions adopted by the EC2 are as follows.

2.2.6.1. *Resistance to compression*

The compressive strength of concrete is denoted by *characteristic resistance* classes measured on a 15 × 30 cylinder ($f_{ck,cyl}$) or a 15 × 15 cube ($f_{ck,cube}$) in accordance with EN 206.

The EC2 is intended for concretes for which f_{ck} is less than 90 MPa.

– For any *t* and untreated concretes, thus:

– $f_{ck}(t) = f_{cm}(t) - 8$ (MPa) for $3 < t < 28$ days;

– $f_{ck}(t) = f_{ck}$ (MPa) for $t \geq 28$ days.

– For treated concrete (thermal, curing, etc.), the resistance depends on the type of cement used and is given by:

$$f_{cm}(t) = \beta_{cc} \cdot f_{cm}$$

– $f_{cm}(t)$ is the average compressive strength of the concrete at the age of *t* days;

– f_{cm} is the average compressive strength of the concrete at 28 days, in accordance with Table 3.1 of the EN 1992-1-1.

– $\beta_{cc}(t)$ is a coefficient that depends on age *t* of the concrete;

– t is the age of the concrete, in days;

– s is a coefficient that depends on the type of cement:

= 0.20 for cements of resistance class CEM 42.5 R, CEM 52.5 N and CEM 52.5 R (class R);

= 0.25 for cements of resistance class CEM 32.5 R, CEM 42.5 N (class N);

= 0.38 for cements of resistance class CEM 32.5 N (class S).

NOTE.– exp{ } has the same value as $e^{()}$.

$$\beta_{cc} = \exp\left\{ s\left[1 - \left(\frac{28}{t}\right)^{1/2}\right]\right\}$$

2.2.6.2. *Tensile strength*

The EC2 differentiates between axial traction and flexural traction. The tensile strength of concrete is defined by three types of tests:

– the direct traction test, which defines the value f_{ctk} (characteristic tensile strength) and f_{ctm} (average resistance);

– the splitting tensile test, which defines the value $f_{ct,sp}$;

– the bending tensile test, which defines the value $f_{ct,fl}$.

2.2.6.2.1. Axial traction

$$f_{ct} = 0.9\, f_{ct,sp}$$

where:

– $f_{ct,sp}$ is the value of the splitting tensile test such that:

- $f_{ctm} = 0.30\,(f_{ck})^{2/3}$ for $f_{ck} < C50/60$;
- $f_{ctm} = 2.12 \ln(1 + f_{cm}/10)$ for $f_{ck} \geq C50/60$.

We get: $f_{ctk} = 0.7 f_{ctm}$ (fractile at 5%)

$f_{ctk} = 1.3 f_{ctm}$ (fractile at 95%).

2.2.6.2.2. Bending traction

The recommended value is:

$$f_{ctm,fl} = \text{Max}\{(1.6 - h/100) f_{ctm}; f_{ctm}\}$$

$$f_{ctk,fl} = \text{Max}\{(1.6 - h/100) f_{ctk}; f_{ctk}\}$$

where f_{ctm} and f_{ctk} designate the previous direct traction and height h of the element in mm.

– Evolution of the tensile strength:

$f_{ctm}(t) = f_{ctm}.\{\beta_{cc}(t)\}^a$, where $a = 1$ if $t < 28$ days

$a = 2/3$ otherwise.

NOTE.– The average flexural tensile strength increases as the height of the element decreases from 600 mm.

For example:

$f_{ctm,fl}= 1.4\ f_{ctm}$ for h = 200 mm

$f_{ctm,fl}= 1.2\ f_{ctm}$ for h = 400 mm

$f_{ctm,fl}= 1.0\ f_{ctm}$ for h = 600 mm.

2.2.6.3. *Deformation of concrete*

– *Elastic deformation*: This depends on the composition of the concrete. The average secant modulus is defined by:

$$E_{cm}= 22(f_{cm}/10)^{0.3}\ (f_{cm} \text{ in MPa})$$

The evolution of the modulus over time is given by:

$$E_{cm}(t) = E_{cm}(f_{cm}(t)/f_{cm})^{0.3}$$

The EC2 also defines a tangent modulus E_c = 1.05 E_{cm} for evaluating deformations and taking creep into account.

NOTE.– This modulus that makes it possible to evaluate the instantaneous deformation is less than the value of *Ei* in the BAEL. However, the EC2 does not calculate deformations from the modulus but rather from the curve.

2.2.6.4. *Shortening by shrinkage and creep*

In section 3: materials of the EC2, the creep coefficients $\varphi(t,t_o)$ are defined as a function of *t*, E_c (tangent modulus) and the class of concrete.

Creep deformation at infinite instant *t* under constant stress σ_c applied at age t_o is given by:

$$\varepsilon_{cc}(t,t_o) = \varphi(t,t_o)\sigma_c/Ec$$

1) For $\sigma_c \leq 0.45\ f_{ck}(t_o)$

The following curves can then be applied:

The creep coefficient values are given in the EC2 as a function of the mechanical strength of the concrete of the height of the part at time t_0 at the end of which the load is applied.

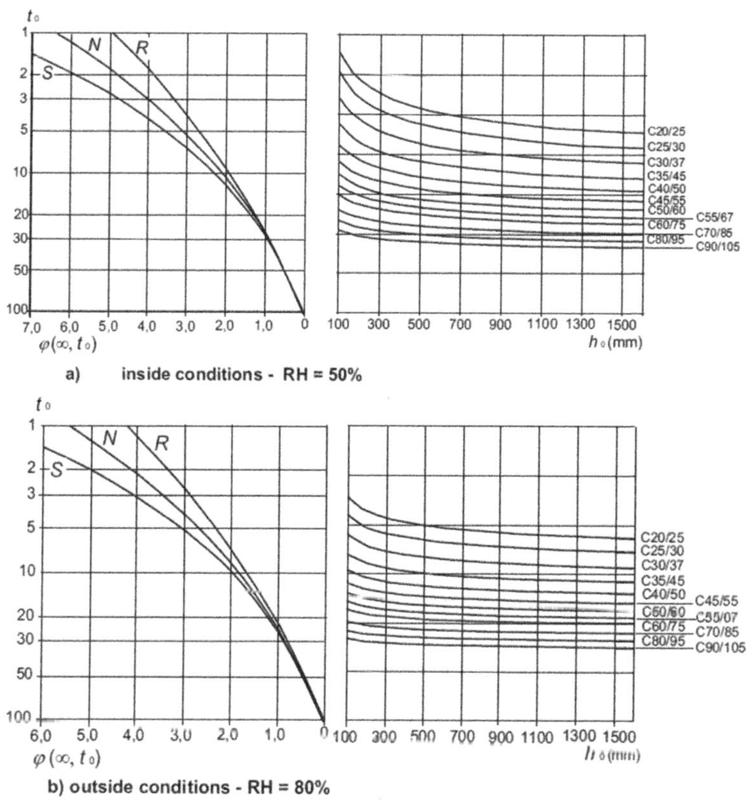

Figure 2.2. *Creep coefficient for concrete under normal environment conditions*

2) *For stress values above 0.45 f_{ck}*, the nonlinear character of the creep should be taken into account.

The formulation recommended by the EC2 is then:

$$\Phi k(t,t_o) = \varphi(t,t_o)\exp(1.5(k_o - 0.45))$$

$$k_o = \sigma_c/f_{cm}(t_o),$$

where t_o is the age of concrete at time of loading.

2.2.6.5. *Shrinking of concrete*

The shrinkage deformation consists of two terms, a shrinkage deformation by desiccation and an endogenous shrinkage deformation.

Let: $\varepsilon_{cs}=\varepsilon_{cd}+\varepsilon_{ca}$

– Uncontrolled desiccation shrinkage values are defined in Appendix B of the EC2 through the following formula:

$$\varepsilon_{Cd,0} = 0.85\left[(220+110\cdot\alpha_{ds1})\cdot\exp\left(-\alpha_{ds2}\cdot\frac{f_{cm}}{f_{cmo}}\right)\cdot 10^{-6}\beta_{RH}\right]$$

$$\beta_{RH} = 1.55\left[1-\left(\frac{RH}{RH_0}\right)^3\right]$$

where:

– f_{cm} is the mean compressive strength (MPa);

– f_{cmo} is 10 MPa;

– α_{ds1} is a coefficient that depends on the type of cement (see section 3 :Materials from EN 1992-1-1).

= 3 for class *S* cements;

= 4 for class *N* cements;

= 6 for class *R* cements;

– α_{ds2} is a coefficient that depends on the type of cement:

= 0.13 for class *S* cements;

= 0.12 for class *N* cements;

= 0.11 for class *R* cements;

– *RH* is the relative humidity of the ambient environment in %;

– RH_0= 100%.

The expression of endogenous shrinkage is given in the EC2 by the following equation:

$$\varepsilon_{Ca}(t) = \beta_{as}(t)\varepsilon_{Ca}(\infty)$$

where:

$$\varepsilon_{Ca}(\infty) = 2.5(f_{ck}-10)10^{-6}$$

and

$$\beta_{as}(t) = 1 - \exp(-0.2t^{0.5})$$

where t is given in days.

2.2.6.6. *Special case of embedded shrinkage (clamping) as defined in EN 1992-3*

– the phenomenon of shrinkage can also occur on elements cast in contact with other concrete structures that have already done a part of their shrinkage and hence clamp the shortening of the new element;

– this results in horizontal, oblique or vertical cracks that will require additional reinforcement to respect the maximum permitted opening w_k;

– the clamping factors are defined in the following elements:

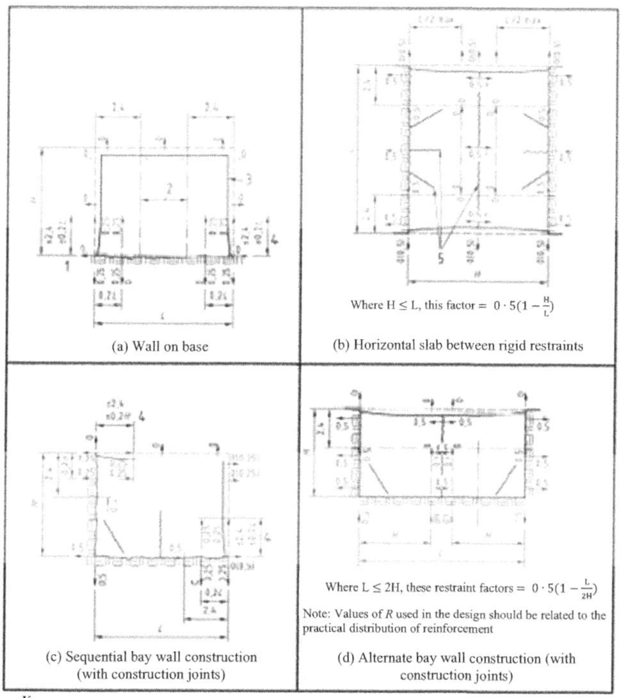

(a) Wall on base

(b) Horizontal slab between rigid restraints

(c) Sequential bay wall construction (with construction joints)

(d) Alternate bay wall construction (with construction joints)

Key
1 Vertical restraint factors
2 Horizontal restraint factor (obtain from table L.1 for this central zone)
3 Expansion or free contraction joints
4 (whichever is the greater)
5 Potential primary cracks

Figure 2.3. *Restraint factors for typical situations*

2.2.6.7. *The deformation of concrete*

– *Transverse deformation*: this is characterized by the Poisson coefficient.

In the EC2, its value is defined by:

– 0 for cracked concrete;

– 0.2 for non-cracked concrete.

– *Longitudinal deformation*: as a result, the total deformation of a component is given by the sum of the instantaneous and deferred deformations:

$$\varepsilon_t = \varepsilon_{ci} + \varepsilon_{cc} = \frac{\varphi}{E_c} \times \sigma + \frac{\sigma}{E_{cm}} = \frac{\sigma}{E_{CM}}(1 + \frac{\varphi}{1.05})$$

All the above elements are summarized in Table 2.1 (from EN 1992-1-1).

Strength of classes															Analytical relation
f_{ck} (MPa)	12	16	20	25	30	35	40	45	50	55	60	70	80	90	
$f_{ck,cube}$ (MPa)	15	20	25	30	37	45	50	55	80	67	75	85	95	105	2.8
f_{cm} (MPa)	20	24	28	33	38	43	48	53	58	63	68	78	88	98	$f_{cm}=f_{ck}+8$(MPa)
f_{ctm} (MPa)	1.6	1.9	2.2	2.6	2.9	3.2	3.5	3.8	4.1	4.2	4.4	4.6	4.8	5.0	$f_{ctm} = 0.3\times f_{ck}^{(2/3)} \leq$C50/60 $f_{ctm}=2.12-\ln(1+(f_{cm}/10))$ >C50/60
$f_{ctk,0.05}$ (MPa)	1.1	1.3	1.5	1.8	2.0	2.2	2.5	2.7	2.9	3.0	3.1	3.2	3.4	3.5	$f_{ctk,0.05}=0.7\times f_{ctm}$ 5% fractile
$f_{ctk,0.95}$ (MPa)	2.0	2.5	2.9	3.3	3.8	4.2	4.6	4.9	5.3	5.5	5.7	6.0	6.3	6.6	$f_{ctk,0.95}=1.3\times f_{ctm}$ 95% fractile
E_{cm} (GPa)	27	29	30	31	33	34	35	36	37	38	39	41	42	44	$E_{cm}=22[(f_{cm})/10]^{0.3}$ (f_{cm} in MPA)
ε_{c1} (‰)	1.8	1.9	2.0	2.1	2.2	2.25	2.3	2.4	2.45	2.5	2.6	2.7	2.8	2.8	see Figure 3.2 ε_{c1}(‰)$=0.7f_{cm}^{0.31} \leq 2.8$
ε_{cu1} (‰)	3.5									3.2	3.0	2.8	2.8	2.8	see Figure 3.2 for $f_{ck} \geq 50$ MPa (ε_{cu1}(‰) $=2.8+27[(98-f_{cm}/100]^4$
ε_{c2} (‰)	2.0									2.2	2.3	2.4	2.5	2.6	see Figure 3.3 for $f_{ck} \geq 50$ MPa ε_{c2}(‰)$=2.0+0.085(f_{ck}-50)^{0.53}$
ε_{cu2} (‰)	3.5									3.1	2.9	2.7	2.6	2.6	see Figure 3.3 for $f_{ck} \geq 50$ MPa ε_{cu2}(‰)$=2.6+35[(90-f_{ck})/100]^4$
n	2.0									1.75	1.6	1.45	1.4	1.4	for $f_{ck} \geq 50$ MPa $n=1.4+23.4[(90-f_{ck})/100]^4$
ε_{c3} (‰)	1.75									1.8	1.9	2.0	2.2	2.3	see Figure 3.4 for $f_{ck} \geq 50$ MPa ε_{c3}(‰)$=1.75+0.05[(f_{ck}-50)/40]$
ε_{cu3} (‰)	3.5									3.1	2.9	2.7	2.6	2.6	see Figure 3.4 for $f_{ck} \geq 50$ MPa ε_{cu3}(‰)$=2.6+35[(f_{ck}-50)/40]$

Table 2.1. *Strength and deformation characteristics for concrete*

2.3. Steels

2.3.1. *The mechanical properties of steels*

Steel is an alloy of carbon and iron, and is the most used metal in civil engineering.

Steels are generally classified depending on their chemical composition and mainly from their carbon content.

Steels

0 0.1 0.25 0.4 0.6 *1* *1.7*

x x x x x x

Extra soft Soft Semi-soft Semi-hard Hard Extra-hard

The carbon content makes it possible to vary the mechanical properties (strength, hardness, elongation).

Thermal, thermomechanical and mechanical treatments as well as the addition of alloying elements can also result in changes in the mechanical properties due to the different structural transformations.

In order to characterize steel, three tests are generally available:

– *The traction test*: this is the most important and the most practiced test; it allows us to determine the properties that can be used in calculations;

– *The hardness test*: this is mainly used in the mechanical industry. It gives information on the hardness– tensile strength relationship;

– *The resilience test*: this is a dynamic test that makes it possible to characterize the "fragility" of the material.

Previously, we saw (section 2.2.5.2 from section 2.2.5 mechanical properties) that, due to its constitution, concrete could not withstand high tensile forces. The advantage of reinforced concrete is therefore that the tensile stresses are taken over by the steel.

Concrete then has the role of transmitting the forces of steels.

Let us suppose that the concrete is not cracked, then the deformation of steel and concrete is identical.

This implies that:

$(\Delta l/l)_{concrete} = (\Delta l/l)_{steel}$

Or $(\Delta l/l)_{steel} = \sigma_b / E_b$

For concrete at 150 kg/cm² of resistance, this gives:

$(\Delta l/l)_{concrete} = 150/225\ _{steel} = 2/3 \times 10^{-3}$ m

Or $(\Delta l/l)_{concrete} = (\Delta l/l)_{steel} = \sigma_a / E_a \Longrightarrow \sigma_a = E_a (\Delta l/l)_{concrete^2}$

$= 2/3.10^{-3} * 2.1.10^6$

$= 1400$ daN/cm²

So $\sigma_a <<$ tensile strength of steel

Let us now suppose that the concrete is cracked; then the transmission of the tensile force between the two concrete blocks is done by the steel.

The tensile stress must then be less than the elastic limit of the steel (elasticity calculation).

2.3.1.1. *Round or smooth round concrete steels*

Until about 1950, these were the only steels used in reinforced concrete. Nowadays, they only constitute 10% of the steels used. They are mainly used as a standby bar because they can be folded and unfolded multiple times without risk of rupture.

We then discern the following:

– *FeE22 mild steels*: these come from the recovery of structural steel sections. The elastic limit is close to 21 daN/mm². They can be bent at 90° cold on a 5 Φ chuck;

– *FeE24 mild steels*: quality metal construction. They were often used in older works. Their elastic limit is 24 kg /mm². They can be bent at 180° cold on themselves.

2.3.1.2. *High-bonding steels*

These are characterized by surface roughness, generally in the form of a helix, which increases the steel-concrete bonding, as well as providing a higher yield strength than soft steels.

In structures to be studied, we find:

– *TOR steels*: with the following properties:

Diameter	Elastic limit (kg/cm²)	Ductility (%)	Folding (°)	Chuck (Φ)
$\Phi > 20$	4,200	15	180	5
$\Phi > 20$	4,000	15	180	5

– *Caron steels*: the properties are identical to TOR steels with the exception of ductility, which is 14%.

2.3.1.3. *Welded mesh*

The first technical manual on welded mesh appeared in January 1958. The permissible stresses recorded at the time were around 25– 28 kg/mm².

Without any standardization, one would refer to the different producers' catalog.

A first step toward standardization began in 1960 and 1963, when the welded mesh companies had founded the ADETS (Technical Association for the Development of Use of Welded Mesh, from the French: *A*ssociation Technique pour le *D*éveloppement de l'*E*mploi de *T*reillis *S*oudés).

Standard panels, as well as a "practical guide to calculating and using welded mesh in floors", were created. In 1979, the AFNOR standards were announced.

Today, the EC2 defines the properties of steel as follows:

– the yield limit (f_{yk} or $f_{0,2k}$);

– the real limit of elasticity ($f_{y,max}$);

– tensile strength (f_t);

– ductility (ε_{uk});

– the ability to fold;

– bonding characteristics f_R;

– fatigue resistance;

– weldability;

– shear strength.

2.3.1.4. *Mechanical resistance*

Steel is characterized by:

– its yield limit f_{yk}, which corresponds to an elongation of 0.2%;

– its characteristic tensile strength f_{tk}.

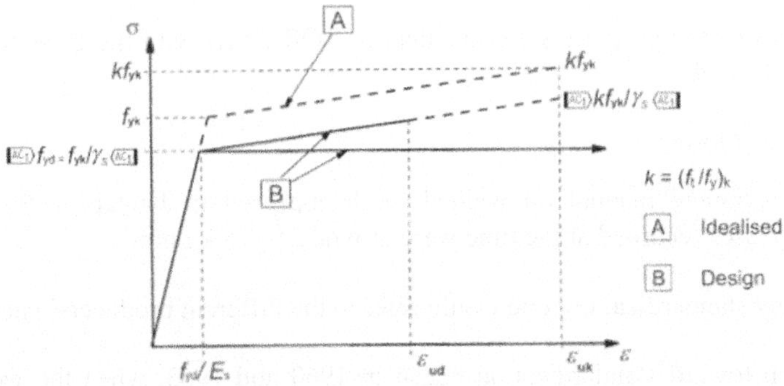

Figure 2.4. *Design stress-strain diagram for reinforcing steel*

The EC2 allows the use of the plastic bearing of the steel. The properties of the steels are summarized in Table 2.2.

Product form	Bars and de-coiled rods			Wire Fabrics			Requirement or quantile value (%)
Class	A	B	C	A	B	C	-
Characteristic yield strength f_{yk} or $f_{0,2k}$ (MPa)	400 to 600						5,0
Minimum value of $k = (f_t/f_y)_k$	≥1,05	≥1,08	≥1,15 <1,35	≥1,05	≥1,08	≥1,15 <1,35	10,0
Characteristic strain at maximum force, ε_{uk} (%)	≥2,5	≥5,0	≥7,5	≥2,5	≥5,0	≥7,5	10,0
Bendability	Bend/Rebend test			-			
Shear strength	-			0,25 Af_{yk} (*A* is area of wire)			Minimum
Maximum deviation from nominal mass (individual bar or wire) (%) — Nominal bar size (mm) ≤ 8 > 8	± 6,0 ± 4,5						5,0

Table 2.2. *Properties of reinforcement*

2.3.2. *Steel-concrete bonding*

2.3.2.1. *Experimental study*

Let us measure the sliding movement of the bar as a function of the force.

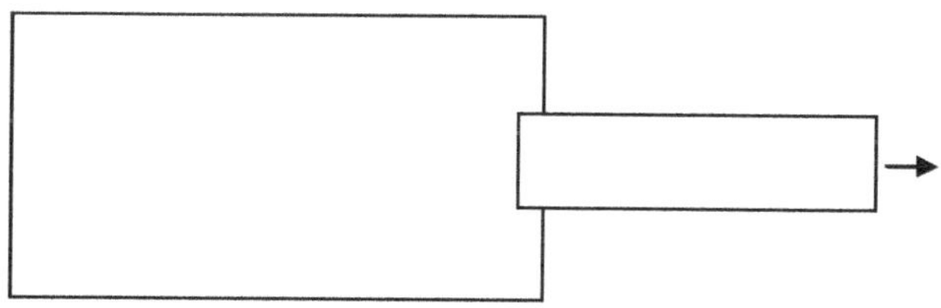

We see that bonding is not a sticking phenomenom, there is always a force to be exerted here.

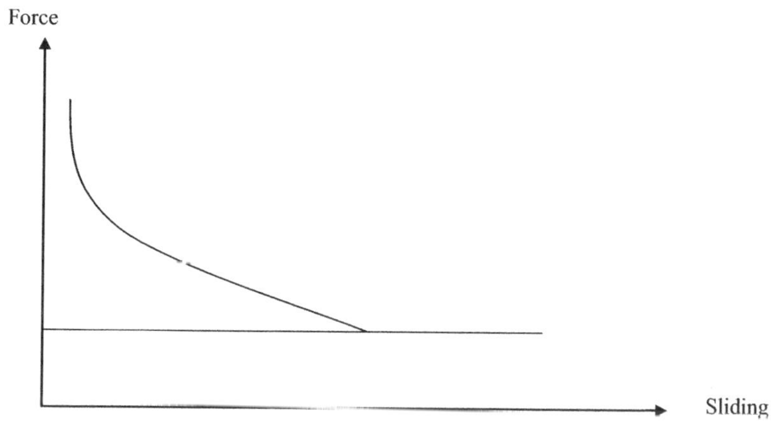

Figure 2.5. *Force-sliding diagram*

The phenomenon of bonding is explained by the fact that the asperities of the bar give rise to compression cones, which oppose the displacement of the bar.

The degree of bonding is defined as the ratio of the tensile strength and the steel-concrete contact surface (= stress).

2.3.2.2. *The two modes of bonding failure*

– a rupture in the support base of the cones. In this case, the bonding rate cancels out.

This failure mode is not permissible in BA. To prevent it, one must:

- place the bars far enough from the edges of the workpiece;

- place transverse reinforcements (stirrups) to oppose any development of the crack;

- a rupture in the cones themselves: the bond rate has a limit value, which is tolerated in BA.

2.3.2.3. *Factors that influence bonding*

– *Roughness of bars and lateral stresses*:

Roughness increases the bond rate, especially rust (subject to brushing the front bars). Once in the concrete, the rusting stops.

The steel must have a set shape.

The higher the bond strength, the higher the compression stress of the concrete sheath that surrounds the bar.

For this reason, steel must be anchored in compressed areas.

– *Traction and repression*:

In both cases, the bonding rate is same.

– *Influence of the transverse reinforcements*:

The bonding is better in the mass than in the vicinity of the walls. Bonding increases with the volume of the transverse reinforcements.

– *Influence of the quantity of concrete*:

The bonding is proportional to the tensile stress of the concrete.

– *Influence of the shape of the bar*:

Circular steels have the best bonding rate.

– *Practical value of the bonding rate*:

The value is between 20 and 40 kg/cm^2.

2.4. Concept of strength of materials

In this chapter, we shall confine ourselves to a quick study on the theory of beams.

In particular, the general assumptions of this theory should be kept in mind:

– the material is assumed to be homogeneous and isotropic. The deformations experienced from external loads are reversible and are very small (theory of linear elasticity);

– displacements of material points between themselves are negligible (so-called 1° order theory).

From these assumptions, the following two laws or principles become apparent:

– *the generalized Hooke law*: this law states that the relations between external forces, stresses and deformations are linear and homogeneous;

– *the superposition principle*: a stress (or deformation) produced by several applied loads is the superposition of the stresses produced by each of the loads that are supposed to act alone.

In the particular case of beams, two other principles are added to these two:

St Venant's principle: the stresses on a section Σ that is far from the points of application of external forces only depend on the stresses of the system that is constituted by forces applied on one side of Σ;

– *the Navier–Bernoulli principle*: when a beam deforms, the cross-sections remain flat.

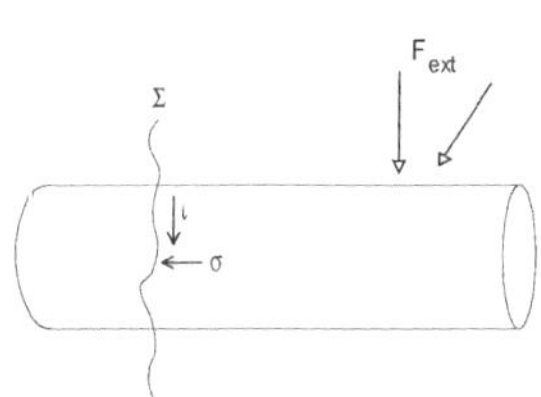

It follows that, in order to apply the theory of beams, it is necessary to ensure that these assumptions are actually respected.

Generally, we consider that reliable results can be derived from this theory if the following conditions are met:

– the width of the beam (transverse dimension) is small relative to its length, in other words:

$\frac{1}{30} < \frac{h}{l} < \frac{1}{5}$ for a straight beam

$\frac{1}{100} < \frac{h}{l} < \frac{1}{5}$ for an arch

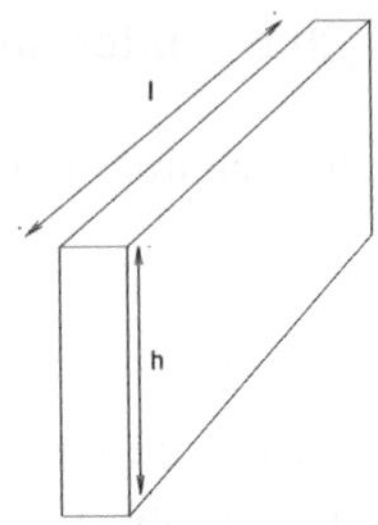

where h is the height of the beam and l is the length of the beam.

– the radius of curvature of the average fiber is greater than five times the height;

– for a variable section beam, the variation must be progressive along the central line of the beam.

The external loads applied to a beam are "actions" that produce "loads" inside the material that forms this beam.

The most common loads are as follows:

– the bending moment (bending of the beam);

– normal force (compression or axial traction);

– shear force (shearing of the beam).

In the following, let us look at the different loads that can be applied to beams.

Notations used:

M = bending moment	t = shear force stress
N = normal force	E = elasticity module
T = shear force	S = section area
σ_e = compression stress	
σ_t = tensile stress	

2.4.1. *Compression/traction*

Let us consider a cross-section of any beam, subjected to an external force that is perpendicular to the latter.

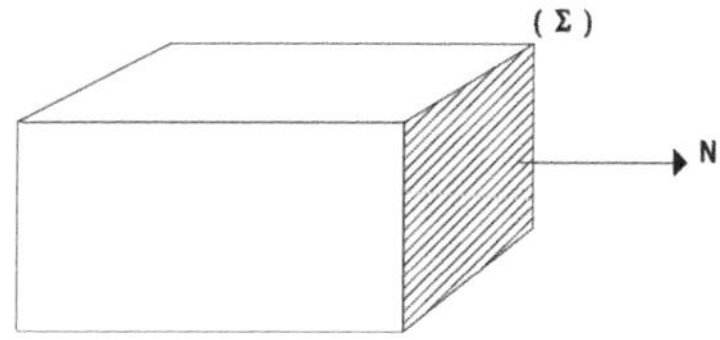

If S is the area of the section (Σ), this force drives a normal stress [perpendicular to (Σ)] on each element of the surface (Σ) and is constant over the entire length of the section. It is equivalent to:

$$\sigma_t = \frac{N}{S}$$

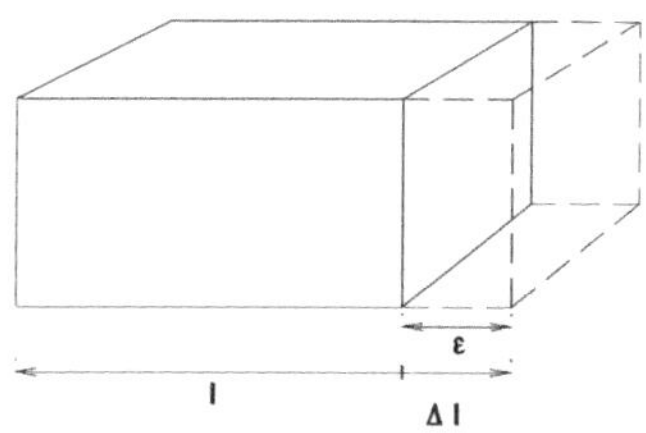

Under the effect of this external force, the fibers of initial length l undergo an elongation Δl, such that:

$$\frac{\Delta l}{l} = \frac{N}{ES}$$

Indeed, according to the generalized Hooke law, we know that:

The deformation is: $= \varepsilon_t = \frac{\sigma_t}{E}$ or $\varepsilon_t = \frac{\Delta l}{l}$ and $\sigma_t = \frac{N}{S}$

Hence $\frac{\Delta l}{l} = \frac{N}{S}$

The deformation therefore acts in the direction of an elongation for a tensile effect, and in the direction of shortening for a compressive force.

2.4.2. *Pure flexion*

Let us consider a beam and two cross-sections of this beam.

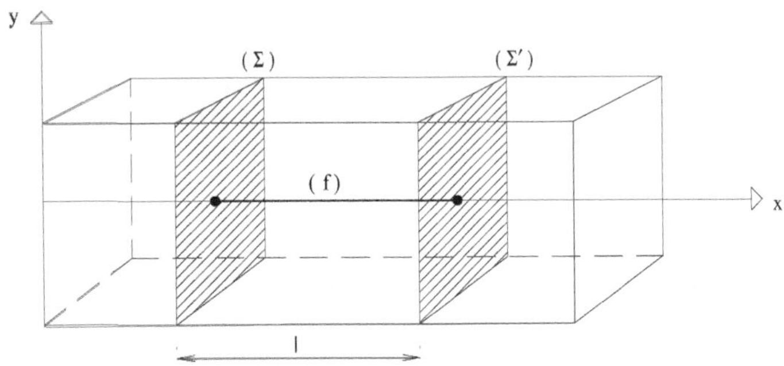

If this beam is subjected to a system of forces that cause the creation of a bending moment, according to the Navier– Bernoulli principle, the sections (Σ) and (Σ′) will remain straight after deformation.

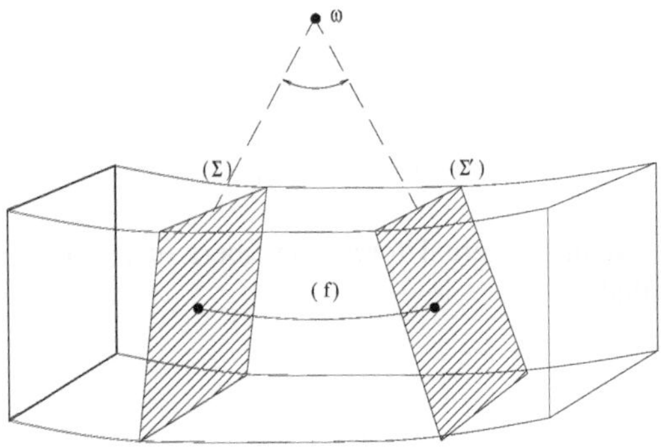

The elongation of any fiber (*f*) between the sections (Σ) and (Σ′) is a linear function of its coordinates in section (Σ).

Taking Hooke's law into account, the stress in the fiber (*f*) is then:

$$\sigma = a + by + cz$$

where the stresses a, b and c are determined by the principle of equivalence:

$$\iint_{(\Sigma)} \sigma \, dy \, dz = 0 \qquad \text{(equilibrium of forces)}$$

$$\iint_{(\Sigma)} \sigma . y . dy . dz = M \qquad \text{(equilibrium of moments)}$$

$$\iint \sigma . z \, dy \, dz = 0 \qquad \text{(M is assumed to be directed according to x)}$$

It ultimately leads to: $\sigma = \frac{M\,y}{I}$

where I is the inertia of the section relative to the main axis that bears the bending moment.

The elongation of the fiber (*f*) can then be considered as:

$$\varepsilon = \frac{\Delta l}{l} = \frac{M.y}{EI}$$

The relative displacement of (Σ) and (Σ′) thus results in a rotation:

$$\omega = \frac{M.L}{EI}$$

The stresses are usually represented as shown in Figure 2.6.

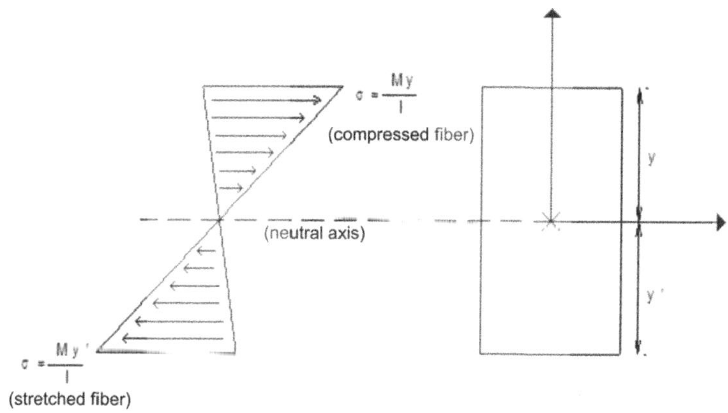

Figure 2.6. *Stresses diagram of a section*

2.4.3. *Shear stress*

Its effect is generally concomitant with that of a bending moment.

The stress of shear force is given by:

$$t = \frac{T_m}{I_b}$$

where T_m is the static moment of the area above the line parallel to Gz.

$$t_{max} = \frac{T}{b_z}$$

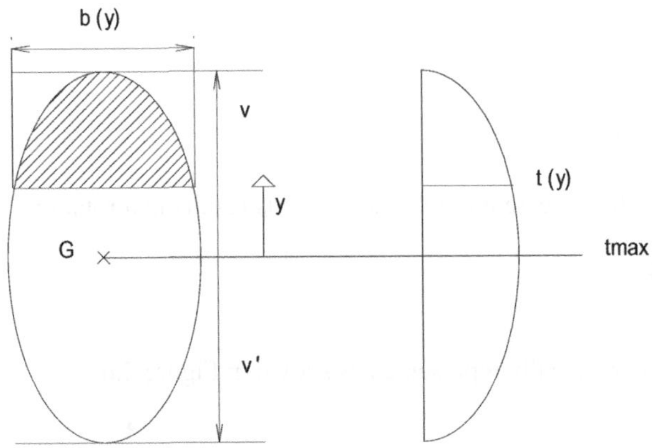

$$z = \frac{I}{\mu_0}$$ = arm of the internal torque lever.

The shear stress deformation is then given as:

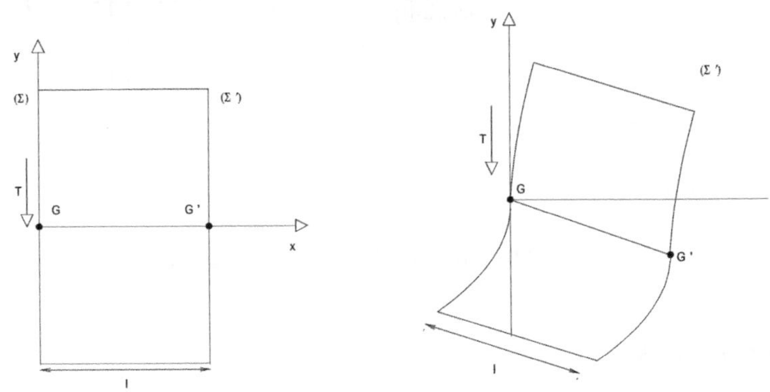

$$\gamma = \frac{T}{GS_I}$$

where S_I is the reduced section
For a rectangle $S_I = 5/6$ S
diamond $S_I = 30/31$ S
circle $S_I = 9/10$ S

2.4.4. *Torsion*

The general torsion problem is relatively complex.

Often, it is studied through the theory of elasticity from which we recall the following elements:

– torsional stresses are tangent stresses that overlap with shear;

– these stresses are perpendicular to the radius vector that comes from the torsion center;

– for a circle, these stresses are proportional to the radius.

2.4.4.1. *Circular section*

When analyzing the deformation of the section due to torsion, point M′ becomes M'_1 and thus:

$$M' M'_1 = \gamma\, dx = \tau/G\, dx = \rho\, d\Theta, \text{ Where } \rho = GM = G'M'$$

Let $\tau = G\rho\, d\Theta/dx$.

The elementary force $\tau\, d\varpi$ then produces an elementary moment $\rho\tau\, d\varpi$.

The sum of these elementary moments must balance the torsion moment Mt, whence:

$$Mt = GI_p\, d\Theta/dx = \tau/\rho\, I_p$$

where I_p is the moment of polar inertia.

From this, we deduce that: $\tau = \rho\, Mt/I_p$ and $d\Theta/dx = Mt/(GI_p)$

2.4.4.2. *For a circle*

$$Ip = Ix + Iy = 2\pi R^4/4 = \pi R^4/2$$

Thus, $\tau_{max} = 0.637\, Mt/R^3$ and $d\Theta/dx = 0.637\, Mt/(GR^4)$.

2.4.4.3. *Rectangular section*

This is studied from the series development and ends up with the same-shaped formula as the circular section:

$d\Theta/dx = Mt/(G*J)$, where J is the torsion modulus $J = Kab^3$.

From this, we deduce that:

$d\Theta/dx = Mt/(GKab^3)$ and $\tau_{max} = K'Mt/(ab^2)$

Caquot approximation for determining parameters K and K′:

Let $m = a/b$

$$1/K = (1 + 1/m^2)[0.225 - 0.035\ ((m - 1)/(m + 1))^2]$$

$$K' = 0.601 - 0.226\ (m - 1)/(m^2 + 1)^{0.5}$$

3

Pathology of Structures

3.1. Pathology of concrete structures

3.1.1. *Cracking*

3.1.1.1. *Definition*

Cracking is the external and visible manifestation of a state of constraint that the material is not capable of tolerating.

This state results from the application of actions that generate stresses leading the material to breaking point.

They can be tensile, compressive or shear stresses that are incompatible with the structure.

3.1.1.2. *Special case for concrete*

By its very nature, concrete has a very low tensile strength. Cracking is therefore usually generated by a redistribution of forces that lead to exhaustion of the tensile strength of the loaded part.

The crack is thus located in the plane on which the main traction stress is exerted.

It also follows that the main traction and compression stresses are exerted on perpendicular faces. These elements are illustrated in the Mohr circle shown in Figure 3.1.

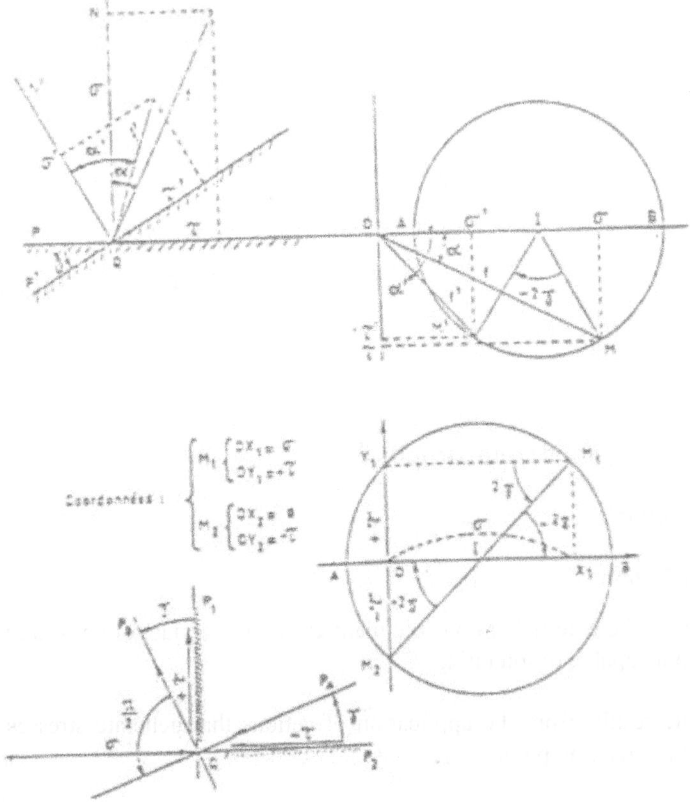

Figure 3.1. *Mohr circles*

APPLICATION.– Using the Mohr circle, the principle follows that a crack from traction is equivalent to a crack from shearing on facets arranged at 45°.

This also explains why cracks from differential settlement are arranged at about 45° and are not purely vertical (Figure 3.2).

3.1.1.3. *Cracking linked to operation of the structure*

These cracks can:

– come from loads that conform to the calculation principle (BAEL weak cracking, for example), which are only dangerous in the case of an inappropriate value for their opening;

– be generated by stresses that do not conform to the calculation scheme. They are then a signal for abnormal operation of the structure.

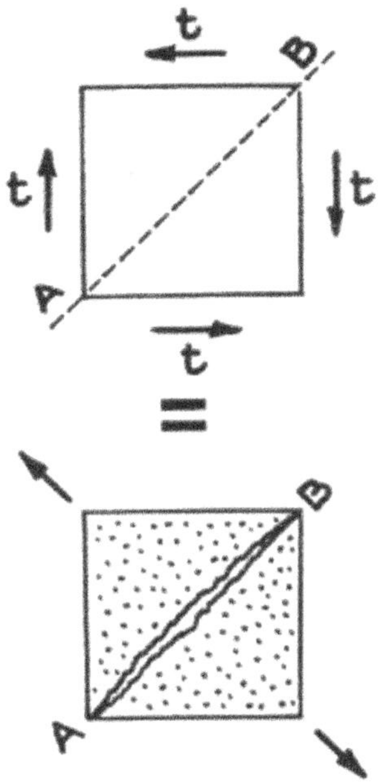

Figure 3.2. *Stresses diagram*

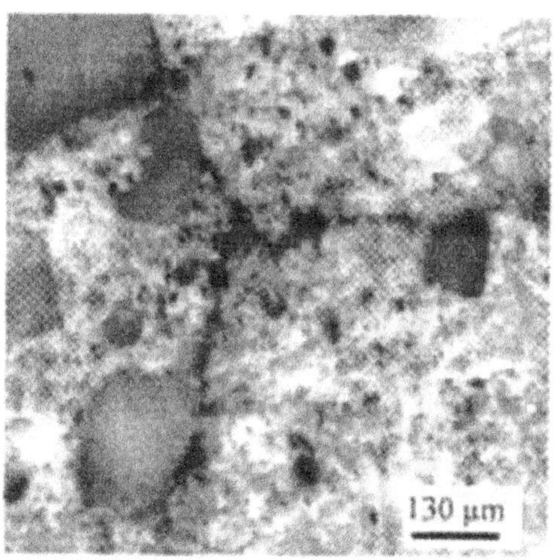

Magnification of image (a) at the level of microcracks

Figure 3.3. *Microcracks inside concrete matrix*

Examples of cracks that occur from operation of the structure:

1) Compression cracks (swelling of the material by Poisson effect):

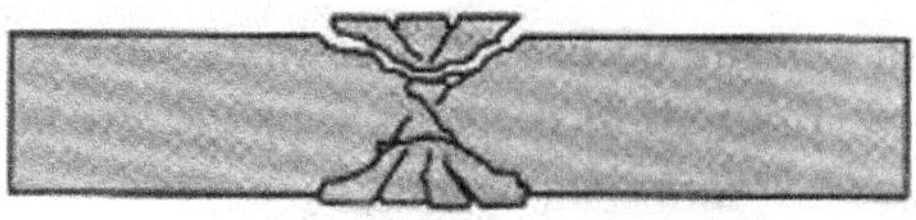

2) Traction cracks (from direct traction):

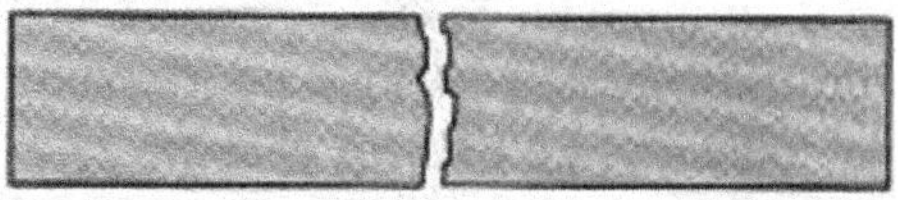

3) Traction cracks (from flexural traction):

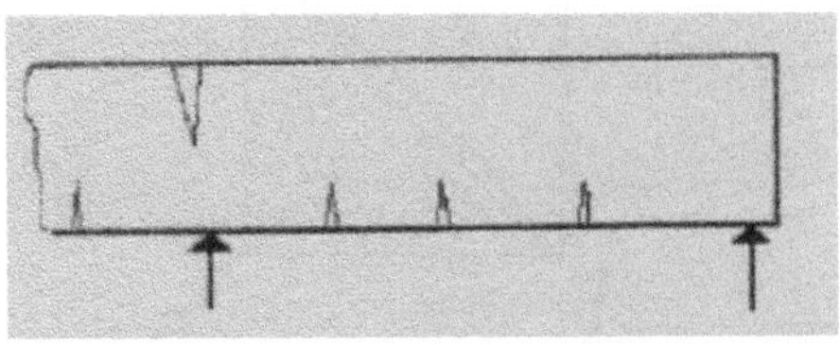

4) Shear cracks (from compression rods on supports):

5) Shear cracks (from torsion):

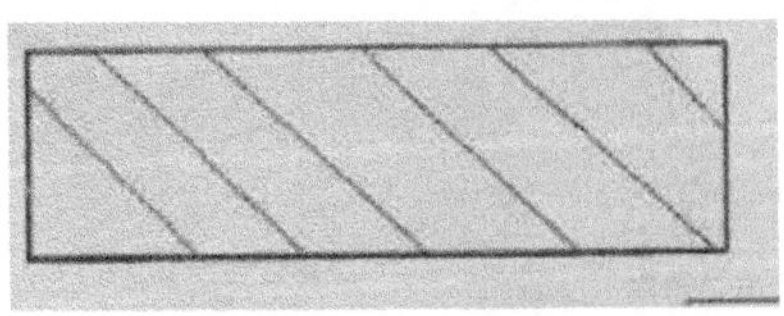

6) Shear cracks (from compression rods along a tie rod):

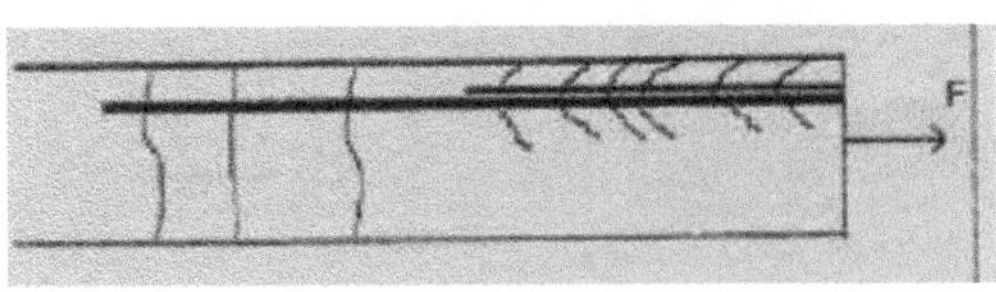

7) Longitudinal cracks along the bars (bonding defect):

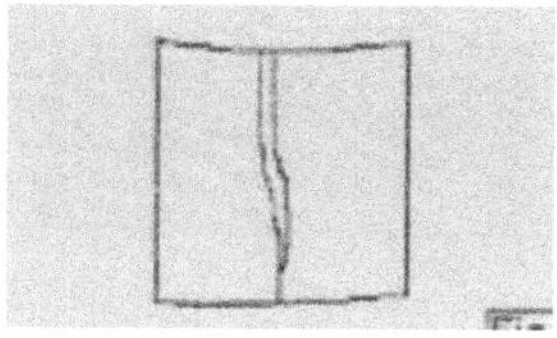

8) Burst cracks in compressed areas of concrete:

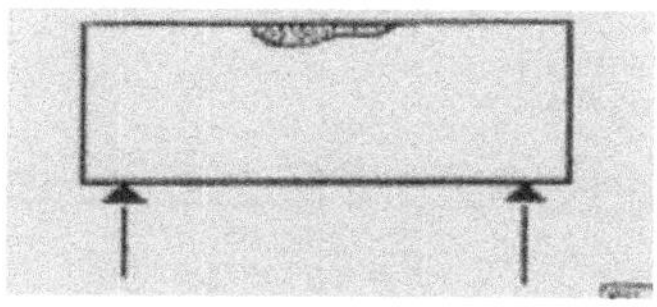

9) Cracks connecting the bars of a beam:

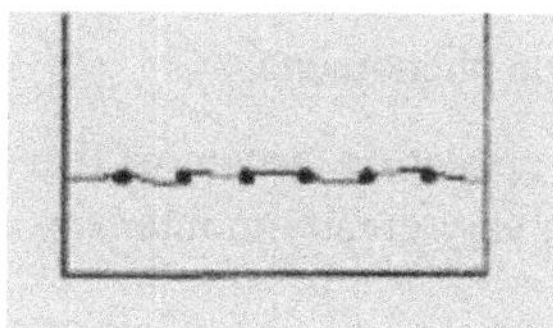

10) Cracks along a key (through the supporting rod and the anchoring area of shear reinforcements):

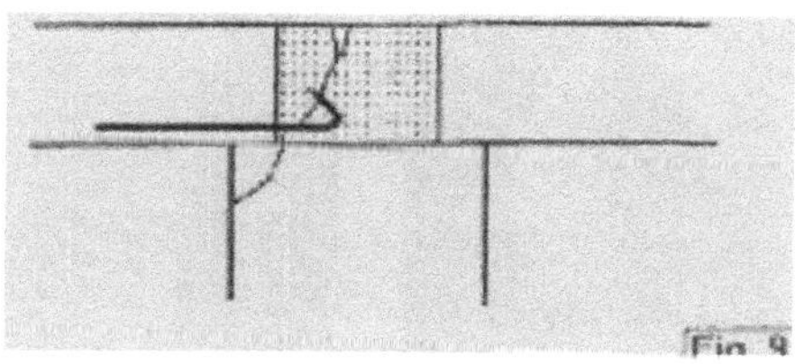

11) Cracks from shrinkage or hygrometry (superficial tensile force following a loss of volume in the material):

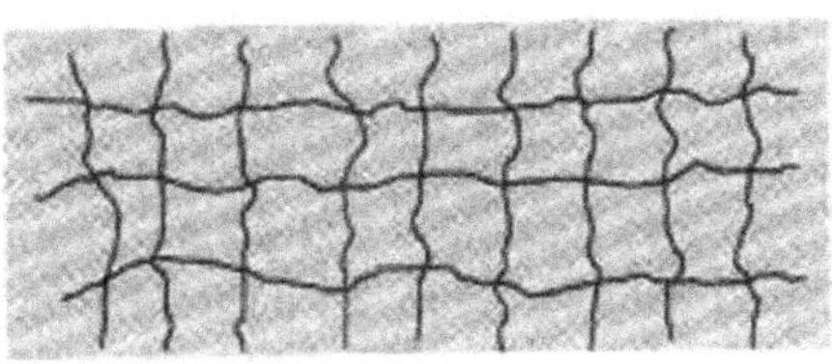

12) Thermal cracks (tensile stress on a restrained beam):

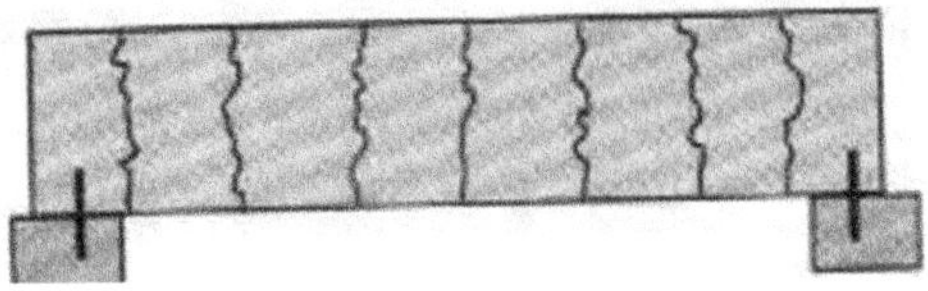

13) Shrinkage cracks (linked to inappropriate construction arrangements):

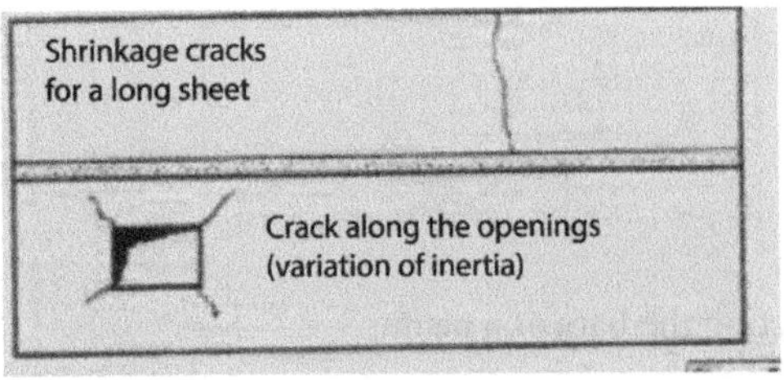

14) Cracks related to fatigue phenomena:

These phenomena tend to appear on structures that have been subjected to loads (static or dynamic), which can vary greatly in intensity and frequency over time.

Tests carried out on BHPs have produced the results seen in Figures 3.4 and 3.5.

Column 3 after breaking
a) Front
b) Rear

Figure 3.4. *Source: Annales ITBTP no. 536*

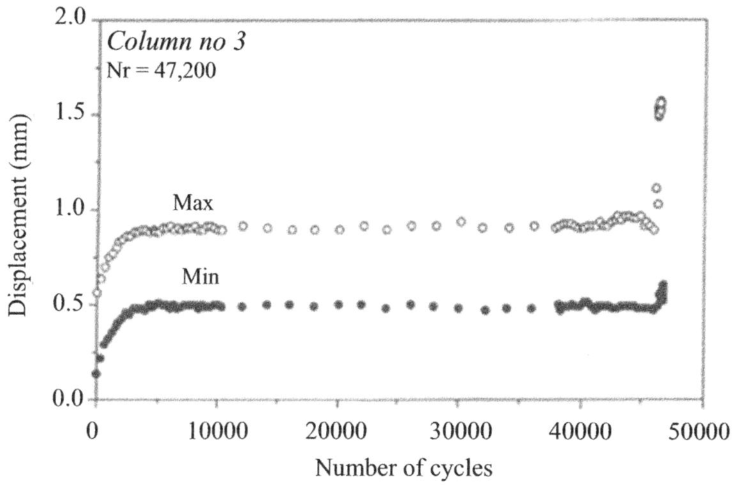

Figure 3.5. *Fatigue tests*

3.1.1.4. *Cracking related to the material and its implementation*

These are summarized in Table 3.1 (adapted from CEB Bulletin 183).

Type of crack	Figure reference	Location	Elements involved	Main cause	Other causes
Plastic placement	A	Below the reinforcement	Thick section	Excessive purging	Fast initial evaporation
	B	On the bars	Top of the columns		
	C	Change in thickness	Slab caisson		
Plastic shrinkage	D	Along the diagonals	Slab	Fast initial evaporation	Insufficient purging
	E	Random	Reinforced concrete slab		
	F	Above the reinforcement	Reinforced concrete slab	Fast initial evaporation, poor cover	
Thermal initial shortening	G	Impaired external deformation	Thick wall	Excessive heat	Fast cooling

	H	Impaired internal deformation	Thick slab	Temperature gradient	
Long-term evaporation shrinkage	I		Slab and thin wall	Ineffective joints	Excessive shrinkage, bad cure
Faïence		Against formwork	Concrete siding	Waterproof formwork	Rich mix Weak cure
	K	Troweled concrete	Slab	Excessive troweling	
Corrosion of reinforcement	L	Natural	Columns and beams	Lack of covering	Low-quality concrete
	M	Calcium chloride	Precast concrete	Calcium chloride excess	
Alkali reaction			Humid environments	Reactive aggregates + cements with alkalis	

Table 3.1. *Different types of cracks*

It should be noted that the first five causes of cracking occur between the first few hours and the first few months of existence of the concrete. Plastic placement is due to the flow of concrete, which can lead to empty pockets under the reinforcements.

Corrosion cracks in reinforcements are the manifestation of an increase in the volume of steels.

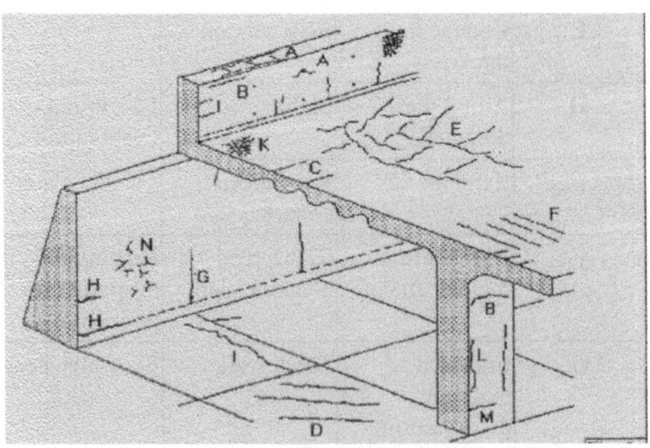

Figure 3.6.

Examples of cracks linked to the material and its implementation

NOTE.– Premature cracking in concrete.

The main causes are as follows:

– "bleeding" water: the appearance of a film of clear water on the horizontal free surface of fresh concrete with gradual settlement of the concrete skeleton under the effect of gravity. Open cracks (sometimes several millimeters) may then appear along the obstacles that oppose this settling movement;

– plastic shrinkage: This is exogenous shrinkage through desiccation that manifests itself before and during the setting;

– thermal contraction after setting: this occurs due to the highly exothermic nature of the hydration reaction. The temperature within the concrete can reach several tens of degrees before returning to normal temperature;

– shrinkage by autodesiccation: this is the isothermal contraction during hydration (endogenous phenomenon);

– restraint (restrained imposed deformation).

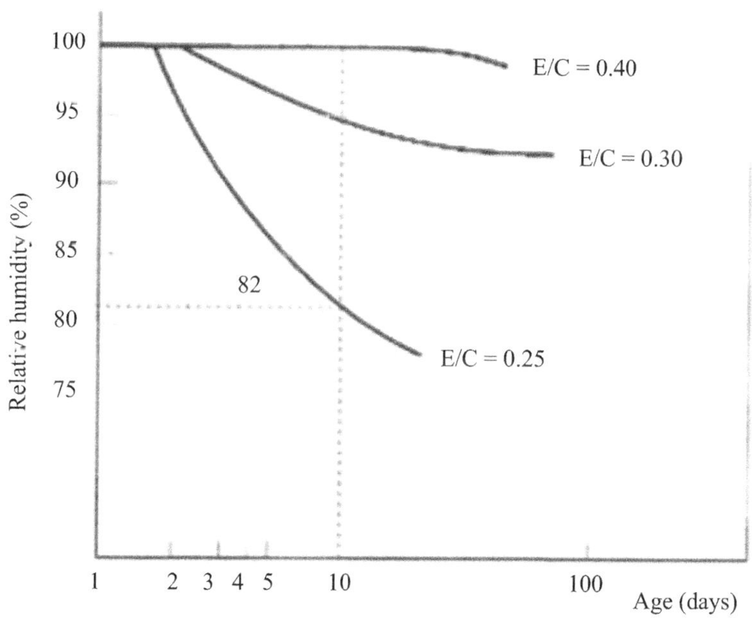

Figure 3.7. *Autodesiccation of the cement paste as a function of the W/C ratio (source LCPC)*

The general causes of premature cracking are summarized in Table 3.2.

Setting phase			
A few hours		A few days	
Time axis			
Causes/ mechanisms	Penetration	Plastic shrinkage	Thermal shrinkage after setting
Van der Waals forces between the fine grains	– Water dosage – Mineralogical nature of fine elements (sand, cement) – Ions in the intersite solution of fresh concrete		
Desiccation through evaporation		– Water dosage – Setting duration (temperature) – Hygrometry/wind speed – Surface/thickness ratio – Distance to the exposed face	
Contraciton through cooling			– Hydration heat curve – Size of the pieces – Insulation of formwork

Table 3.2. *Causes of premature cracking*

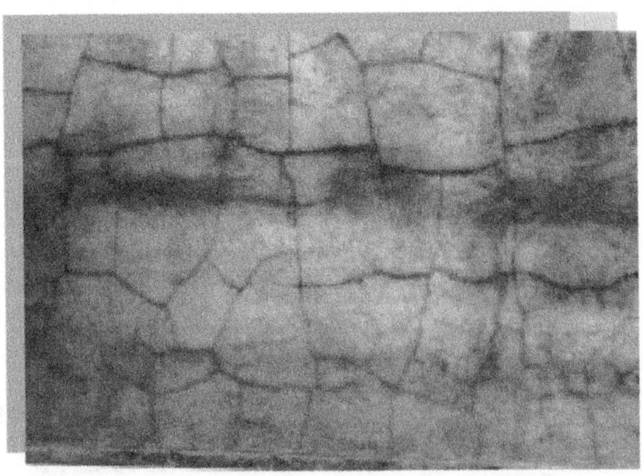

Figure 3.8. *Example of cracking by plastic shrinkage*

In 2015, under the CEOS project (Behavior and Evaluation of Special Structures, from the French: *Comportement et Évaluation des Ouvrages Spéciaux*), which aimed to validate the Eurocode 2 formulas for calculating cracking for special structures and in particular for *solid structures* (footings, flooring, etc.), the "Recommendations for the control of cracking phenomena" were published.

Concerning the effects of hydration of concrete at a young age, the main lessons to take home from this study are the following:

– if a part is restrained during the cooling phase, the resulting tensile stresses may cause cracking. This may be, for example, restraint of a raft from friction against the ground on which it is poured or of the uplift of a sheet that is restrained at its junction with the same raft;

– if the part is free, temperature differences can cause premature cracking of concrete.

Here are the differences in temperature (thermal gradient) that may lead to cracking:

– thermal gradient between the core and the surface of the part during and after a rise in temperature of concrete, demolding or during the curing period. Cracking can occur:

- in the short term: within 3 days of casting. This is usually a crack in the siding in the absence of surface reinforcement or overly fast demolding. In this case, the tension area of the part should be limited to 20% of its thickness on each siding;

- in the long term: within 10–30 days during the cooling of the core, the part may be restrained by the next casting. This restraint can generate a crack in the core of the part, which may open up on the cladding.

– temperature difference between a new concrete uplift and the previous uplift, which is linked to the new casting. It intervenes within 10–30 days after concreting;

– temperature difference between two pieces of concrete of different thickness but cast in a single phase.

The study also highlighted a *scale effect* for solid parts.

In fact, it has been noted that when the loaded volume is large, the tensile stress of the concrete decreases (Weibull scale effect).

This scale effect is all the more significant the lower the quality of the concrete (low compression stress).

3.1.1.5. *Properties of a crack*

A crack is characterized by the following elements:

– age: this is the most difficult parameter to estimate when it is not directly related to a known accidental cause. It is useful for assessing the state of a crack (obstruction by the formation of lime crystals, making it difficult to inject...). We consider that a crack that is less than 2 years old can be easily injectable even if calcification in tanks were observed upon the filling of water;

– opening: this is the maximum value of the distance between the lips. It is easily measurable for rectilinear cracks (fissurometer, thread count, gauge, linear measuring gauge, etc.) but is less so for cracks with random appearances;

– trace: this is the orientation and the measurable length of the crack. The orientation, as we mentioned above, is revealing of the origin of the pathology. If a crack is continuous in its orientation axis, it is called a *free crack*. If the orientation axis is interrupted, it is called a *discontinuous crack*. The length of the crack is commonly considered as the developed part of the visible section;

– depth: a crack is said to be *transversant* if it is visible on both sides of the piece, it is said to be *blind* if it is transversant but cannot be accessed on one of the two faces (for example a semiburied tank for the face on the ground side). A crack is said to be *superficial* if its opening is at its maximum on the surface and is null a few centimeters further;

– activity: this is the ability of a crack to vary dimensionally over time. We distinguish between *dead cracks* (constant opening regardless of the stresses, such as variations in temperature, applied loads, etc.) and *active cracks* (variable openings depending on the external factors such as those just mentioned). The variation of the opening is called the *movement* of the crack.

The following designations are generally accepted:

– microcracks: crack with an opening smaller than 2/10° mm;

– cracks: opening between 2/10° and 20/10° mm;

– lizards: cracks with opening greater than 20/10° mm.

3.1.2. *The degradation of concrete*

Concrete is generally considered to be a durable material provided that the formula is well prepared, properly sized and properly implemented. However, it is a material that remains sensitive to physicochemical, mechanical and sometimes biological actions.

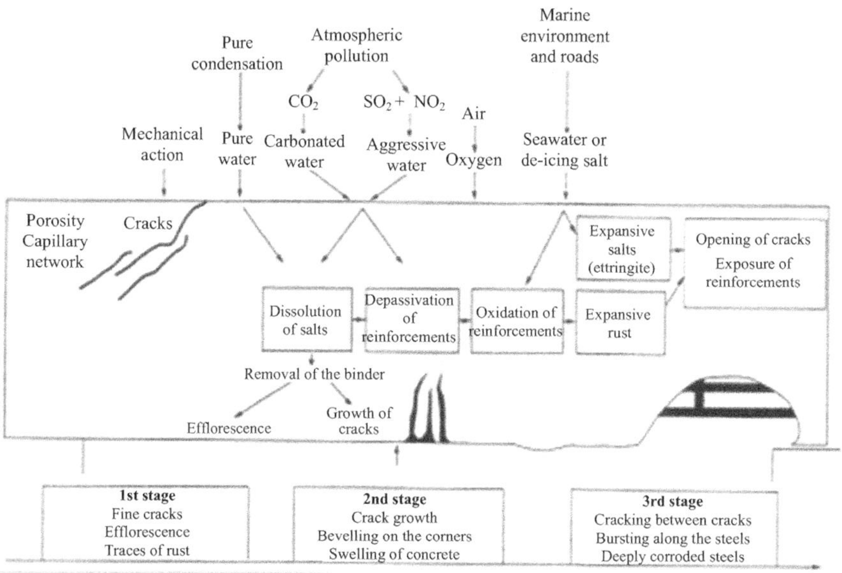

Figure 3.9. *Diagram of the degradation of concrete and reinforcement corrosion*

3.1.2.1. *Physicochemical degradation*

3.1.2.1.1. Carbonation and corrosion of steels

Concrete in tanks and, more generally, in sewage or water-treatment plants is in contact with air, water (or effluents) and soil (or the ground).

However, ambient air contains carbon dioxide (between 0.03 and 0.10%), which, when hydrated (by rainwater, for example), becomes a weak acid (carbonic acid H_2CO_3).

The Portlandite that is present in cement then reacts to form lime carbonate:

$$Ca\,(OH)_2 + CO_2 + H_2O - CaCO_3 + 2H_2O$$

This action has the effect of lowering the pH of the interstitial phase of concrete to make it fall below 9 from an initial value of 13.

Although carbonation is favorable for concrete (it could be compared to the formation of cullet for calcareous stones), it is damaging for the reinforcements, which are at a pH that no longer guarantees their passivation.

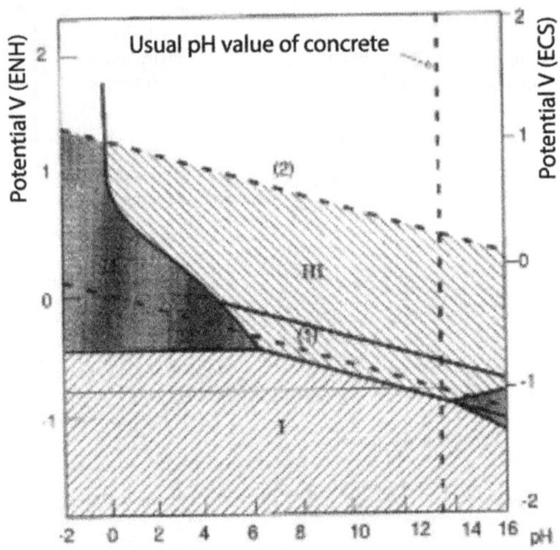

Figure 3.10. *Diagram of the corrosion of steels: Pourbaix diagram for the Fe-H_2O system. Domain I: immunity domain in which iron does not corrode. Domain II: corrosion domain in which Fe^{2+} and $FeOOH^-$ ions are formed. Domain III: passivity domain where iron coats itself in Fe_3O_4 or Fe_2O_3*

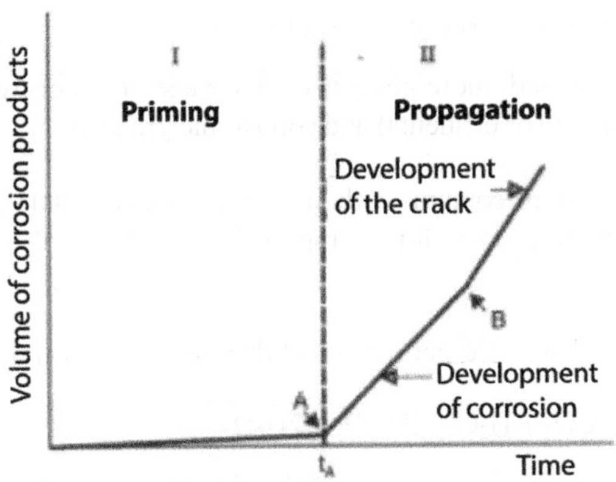

Figure 3.11. *Diagram of the kinetics of the behavior of reinforcements and concrete [TUU 82]*

After a more or less long firing phase, which depends on the coating of reinforcements, the compactness of concrete, etc., there is a rapid development of corrosion and consequently, an expansion effect and the appearance of cracks in the area around the steels.

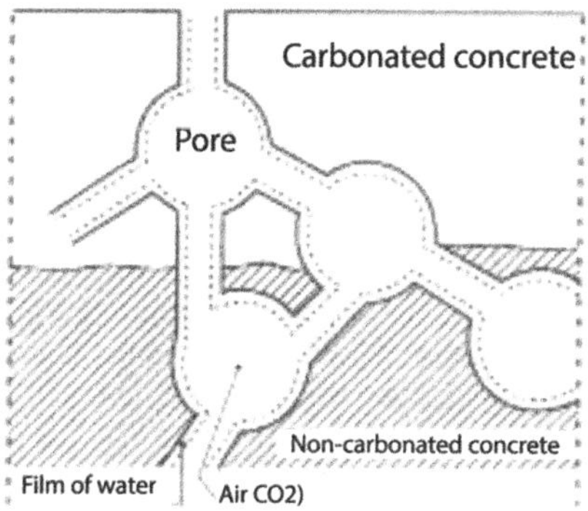

Figure 3.12. *Diagram of carbonation of concrete*

Various tests have been carried out to test the influence of the parameters involved in the composition of concrete or in the exposure of the element to the carbonation depth.

These tests are summarized below:

– influence of the compressive strength of concrete (Figure 3.13).

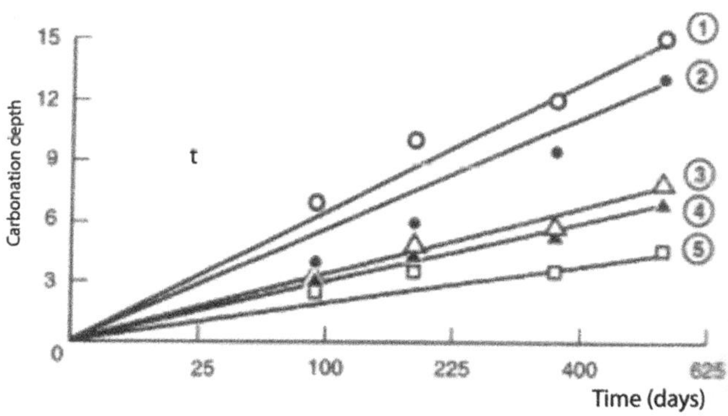

Figure 3.13. *Evolution of the carbonation depth as a function of time [BAL 92]. Curves 1–5: CPJ-CEM II 32.5 concretes with fc_{28} values of 20, 25, 30, 35 and 40 MPa*

It appears that the higher the compressive strength, the slower the carbonation.

– influence of the compressive strength of concrete on the carbonation depth (Figure 3.14).

The carbonation depth can be approximated by the following formula:

$$e = 125 \exp(-0.05 f_{c28})$$

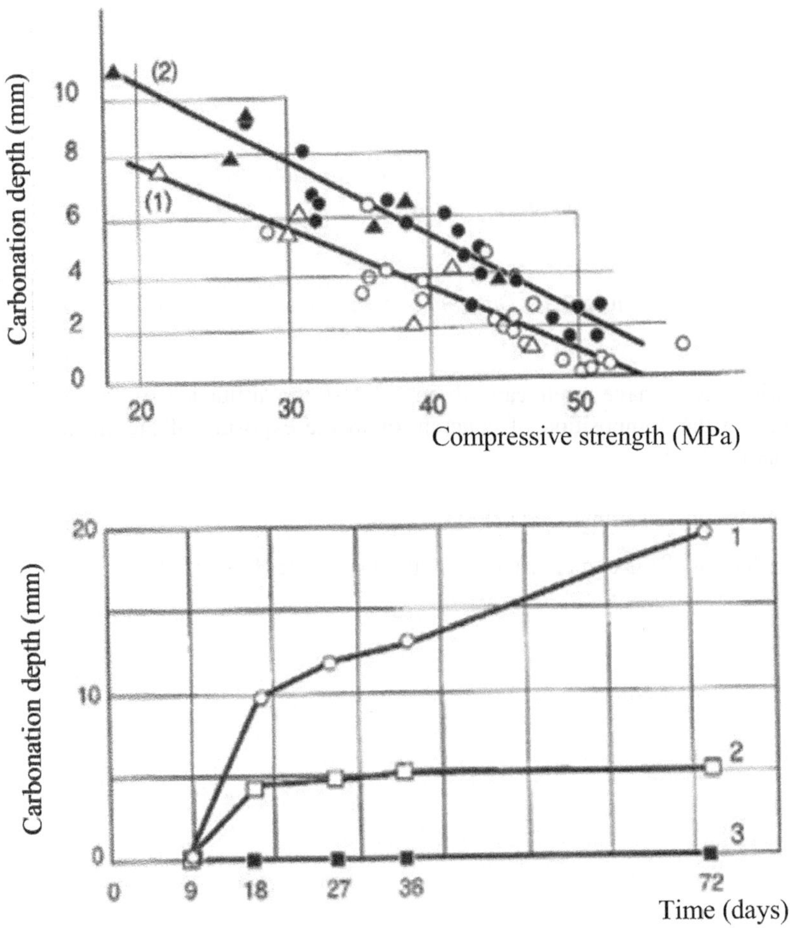

Figure 3.14. *Comparison between the carbonation of ordinary concrete (C25/30) curve 2 and HP concrete (C60/75) curve 1*

– influence of hygrometry from the external environment (Figure 3.15).

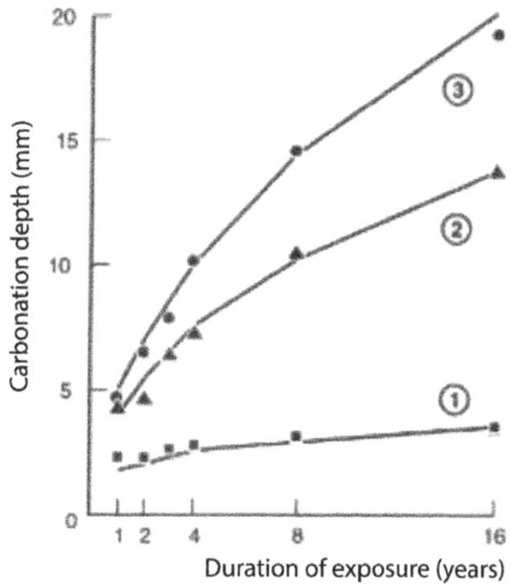

Figure 3.15. *Influence of humidity on the progression of carbonation [WIE 84]. Curve 1: t = 20 °C and 65% RH (external atmosphere). Curve 2: t = 9 °C and 77% RH (external and under cover). Curve 3: t = 9 °C and 77% RH (external atmosphere under rain exposure). This experiment shows that the phenomenon of carbonation develops more deeply in concretes subjected to increased hygrometry than in others*

– influence of the W/C ratio (Figure 3.16).

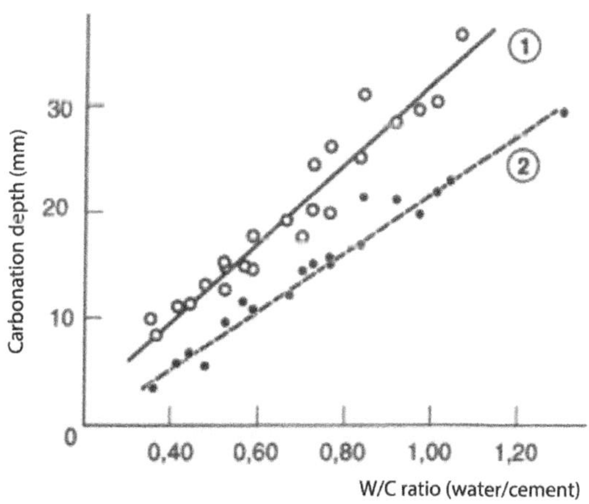

Figure 3.16. *Influence of the W/C ratio on carbonation depth [SKJ 86]. Curve 1: specimen preserved in its mold for 1 day and in water for 27 days. Curve 2: specimen preserved in its mold for 1 day. Carbonation depths are measured after 6 years of exposure*

The influence of the quantity of water is clearly apparent with increasing carbonation depth.

– influence of cement dosage (Figure 3.17).

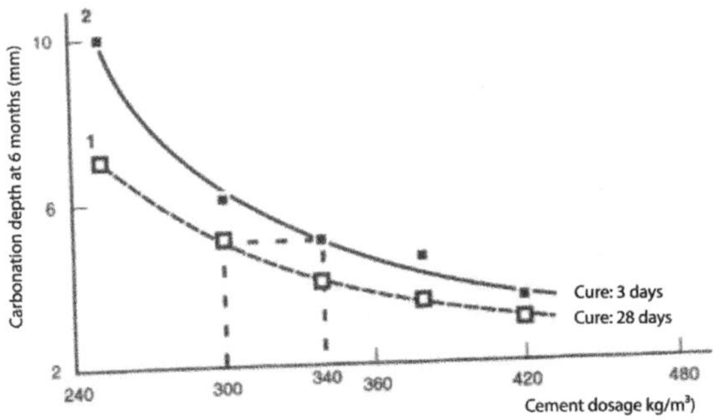

Figure 3.17. *Influence of cement dosage and the curing time on carbonation depth. An increase in cement dosage is favorable for carbonation depth*

In conclusion, the pathology linked to carbonation, which depends on the aforementioned parameters, is mainly subjected to a given base that covers the steels.

The BAEL, FDP 18.011 and EN 206 requirements are as follows:

– for concrete that is in contact with the ground, external sheets, concrete in contact with weakly aggressive liquids (class XA1): minimum cover 3 cm;

– concrete in contact with medium aggressive liquid: 4 cm;

– concrete in contact with brackish water or in tidal zones: 5 cm.

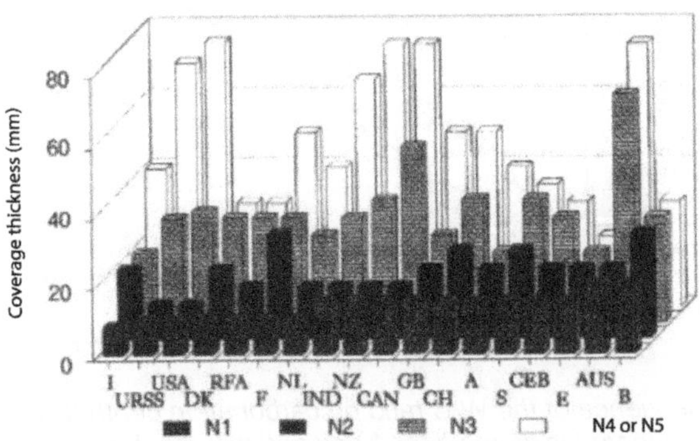

Figure 3.18.

EXAMPLE.– This is the verification of cover thickness in a wastewater plant sandblaster.

Figure 3.19. *The walls after pouring and while undergoing reinforcement*

The results of auscultations performed on the pachometer are as follows (Figure 3.20).

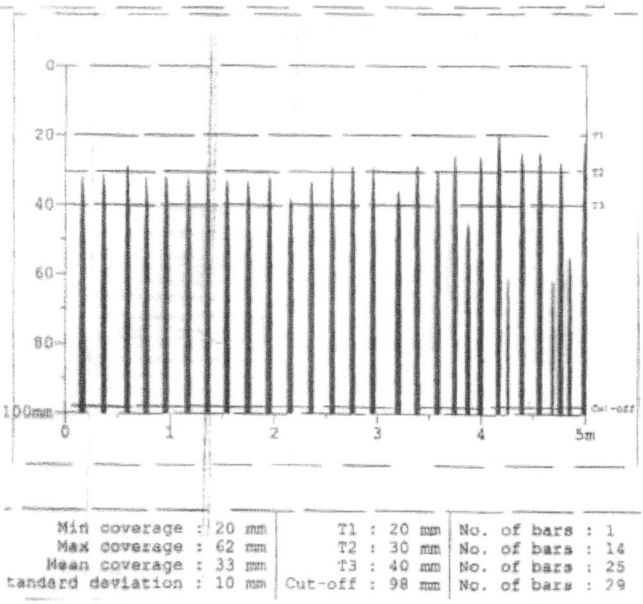

Linear length	(m)	1/5 5	2/5 5	3/5 5	4/5 5	5/5 4.2	Total 24.2
Cover < 20 mm	n	10	1	23	3	5	42
	n_{20}/N	25.0%	2.9%	65.7%	8.8%	16.1%	24.0%
Cover < 30 mm	n	23	15	27	14	26	105
	n_{30}/N	57.5%	42.9%	77.1%	41.2%	83.9%	60.0%
Cover < 40 mm	n	30	31	32	21	28	142
	n_{40}/N	75.0%	88.6%	91.4%	81.8%	90.3%	81.1%
	N	40	35	35	34	31	175
Min cover	(mm)	12	19	14	14	13	12
Max cover	(mm)	89	52	80	63	59	89
Medium cover	(mm)	32	32	24	38	27	30
Standard deviation	(mm)	17	7	13	14	9	

Figure 3.20. *Pachometer measures of cover*

For a regulatory cover of 4 cm, it appears that up to 91.4% of reinforcements comply with this condition.

3.1.2.1.2. Sulfate reactions

These reactions may occur in the marine environment or in effluents loaded with sulfates. In simple terms, sulfate reactions are the result of sulfates attacking the lime and the aluminates in the cement. The final product is ettringite (or Candlot salt). The mechanism is as follows:

$$Ca(OH)_2 + H_2SO_4 - CaSO_4 + 2H_2O$$

$$3CaSO_4 + C3A + 32H_2O - C3A.3CaSO_4.32H_2O\text{–expansion}$$

(Ettringite)

This salt is present in healthy concrete.

There are three types of ettringite that can coexist in the same concrete:

– primary formation ettringite (without expansion);

– secondary formation ettringite (possible expansion);

– delayed ettringite following a rise in temperature (possible expansion).

1) *Primary formation ettringite* is the product of the hydration of cements. The crystals formed (acicular form) appear before the hardening of concrete in the open spaces and thus contribute to good cohesion of the cementate paste (cohesion at young age).

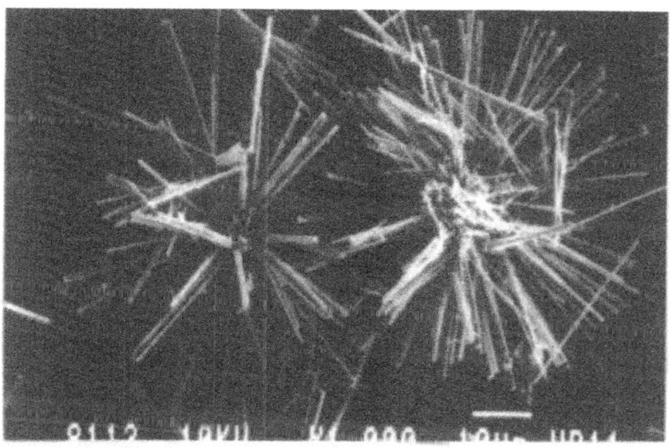

SEM view of the front showing a sea urchin appearance of ettingite needles

Figure 3.21. *Ettringite formation (source: LCPC)*

2) *Secondary formation ettringite* crystallizes in hardened concrete because of circulation of water in concrete and due to sources of external sulfates (soil, etc.) or internal sulfates (overly high quantities in the constituents of concrete).

It crystallizes in acicular form in the free spaces in the concrete (pores, cracks, paste-aggregates interface, etc.). The secondary formation ettringite that results from an external or internal supply of sulfates is likely to generate swellings.

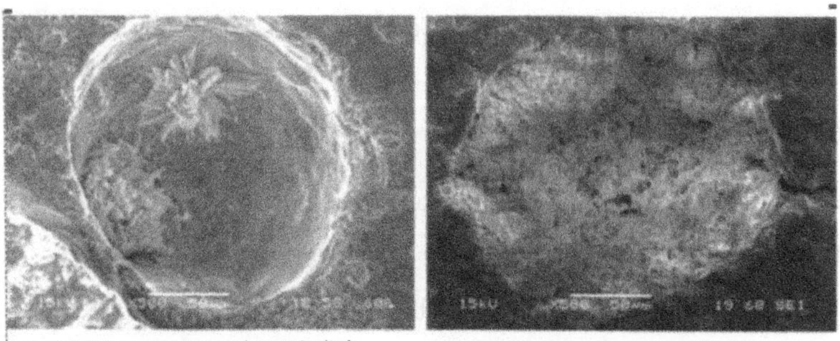

Photo 5: Non-expansive secondary ettringite in a pore of cement paste. SEM, polished section, SEI, x500

Photo 6: Non-expansive secondary ettringite in a pore of cement paste. SEM, polished section, SEI, x500

Figure 3.22. *Ettringite formation*

3) *Delayed ettringite formation* only concerns concretes that have undergone a temperature increase of more than 65–70 °C at a young age.

Crystals may form upon return to ambient temperature and in the presence of moisture; this causes swelling pressures followed by expansion.

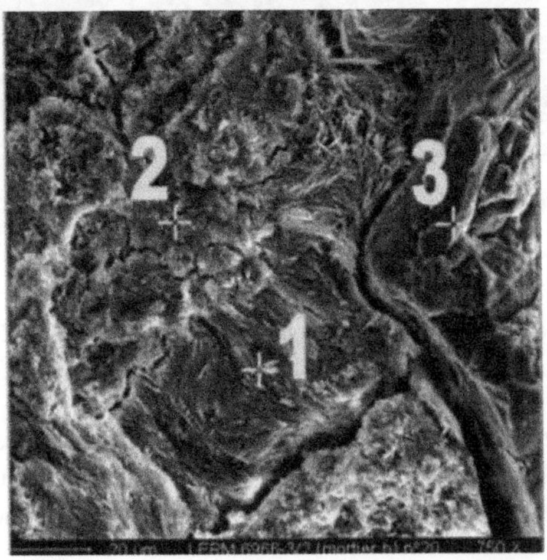

Figure 3.23. *Three types of ettringite (Source: LERM)*

These sulfate reactions mainly occur through:

– actions of selenite waters;

– actions of seawater (containing 2.2 g/L of $MgSO_4$);

– actions of embankments or soils containing sulfates;

– actions of acid rain (causing the SO_2 contained in the atmosphere).

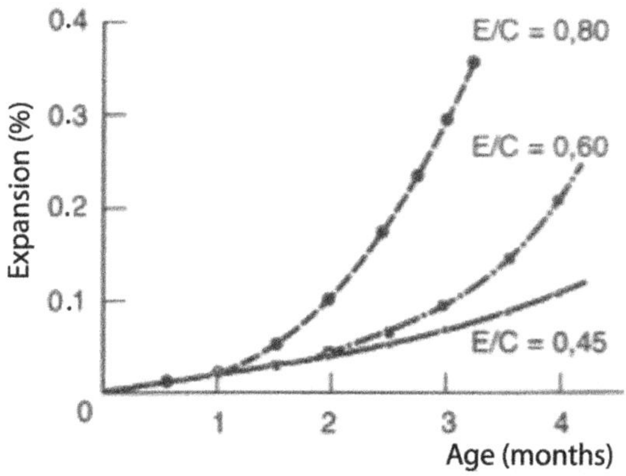

Figure 3.24. *Influence of the W/C ratio on sulfate attacks [OUY 88]*

The aggressiveness of the sulfate environment depends on the concentration of SO_4^{2-} ions but also on the nature of the cation (Ca^{2+}, Mg^{2+}, Na^{2+}, NH_4^+). The FD P 18.011 documentation booklet gives guidance on this.

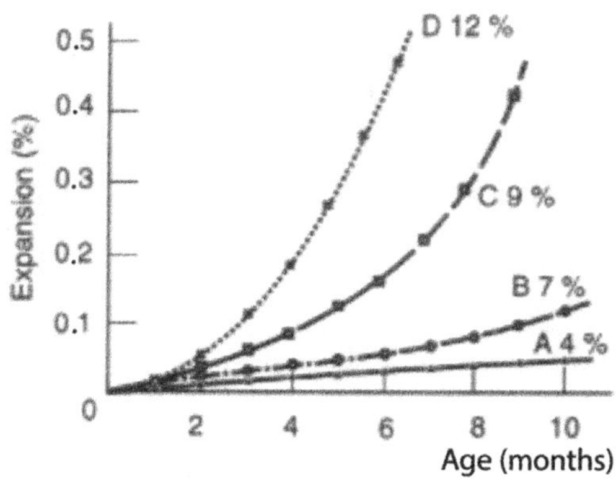

Figure 3.25. *Influence of C_3A content on sulfate attacks*

Cement	MgO	SO_3	C_3A
A	0.8	2.6	4.3
B	1.0	2.5	7.0
C	1.2	3.1	8.8
D	0.9	2.0	12.0

As a result of this study, cements containing tricalcium aluminate are particularly sensitive to sulfate attack.

EN 206 recommends the following (Table 3.3).

Table 2 — Limiting values for exposure classes for chemical attack from natural soil and ground water

Chemical characteristic	Reference test method	XA1	XA2	XA3
Ground water				
Mg^{2+} mg/l	ISO 7980	≥ 300 and ≤ 1 000	> 1 000 and ≤ 3 000	> 3 000 up to saturation
Soil				
SO_4^{2-} mg/kg[a] total	EN 196-2[b]	≥ 2 000 and ≤ 3 000[c]	> 3 000[c] and ≤ 12 000	> 12 000 and ≤ 24 000
Acidity ml/kg	DIN 4030-2	> 200 Baumann Gully	Not encountered in practice	

[a] Clay soils with a permeability below 10^{-5} m/s may be moved into a lower class.

[b] The test method prescribes the extraction of SO_4^{2-} by hydrochloric acid; alternatively, water extraction may be used, if experience is available in the place of use of the concrete.

[c] The 3 000 mg/kg limit shall be reduced to 2 000 mg/kg, where there is a risk of accumulation of sulfate ions in the concrete due to drying and wetting cycles or capillary suction.

Table 3.3. *Limiting values for exposure classes according to EN 206*

Figure 3.26. *Examples of sulfate reactions on a mud tarpaulin (waste water plant)*

EXAMPLE.– Special case for an internal sulfite reaction (ISR).

Within concrete, there is a source of sulfates (cement, water, aggregates) that can create a delayed formation of ettringite and thus degrade the concrete part.

The main cause (which is essential but not sufficient) is the elevation of temperature during the setting of concrete.

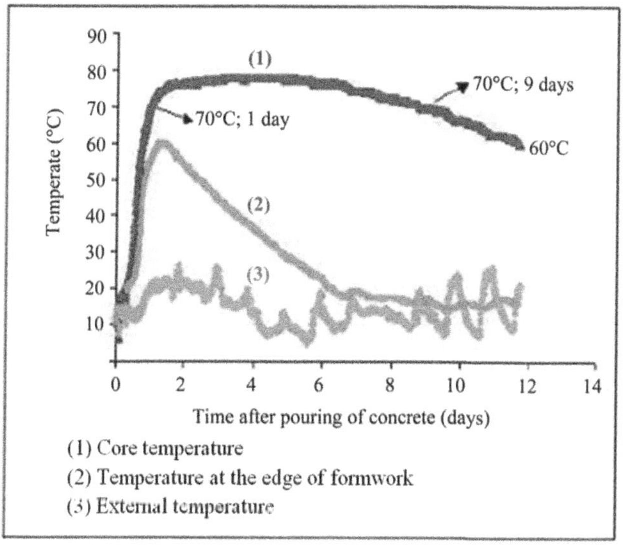

Figure 3.27. *Recording of temperature rises in a solid piece (4 × 5 × 6 m)*

The internal sulfate reaction results in an increase in volume of the part, accompanied by cracking in the concrete surface.

The 2007 LCPC Guide issued a number of recommendations to avoid this phenomenon:

– avoid pouring at very steady rate;

– choose the cement and formula of concretes (LH cement, chilled water, cooled aggregates, etc.). In particular, CEM I 52.5 R cements should be avoided for a solid part;

– avoid high temperatures.

Figure 3.28. *ISR on a river bridge pile*

To identify solid parts is not a straightforward procedure. For example, a 1.50 m thick sole that is injected with C30/37 concrete dosed at 370 kg/m^3 of CEM III/A 42.5 N cement will reach a maximum temperature of 49 °C, whereas a 0.60 m thick cast with C40/50 concrete dosed at 400 kg/m^3 of CEM I 42.5 R cement will reach a core temperature of 65 °C.

The notion of a critical piece has therefore been defined as: "A concrete piece for which the heat is only partially released externally, leading to a significant rise in the temperature of the concrete (thickness above 0.25 m)".

From this, the structure or the piece should be categorized in relation to the risk that one is ready to accept.

This is a choice to be made by the developer and depends on:

– the nature of the structure;

– its end use;

– its impact on safety;

– its aintenance;

– its durability.

The categories are as follows (Table 3.4).

Category	Examples of structures
Category I: Low or acceptable consequences	Structures of resistance classes lower than C16/20 Non-load bearing elements Replaceable parts Temporary structures Non-structural prefabricated products
Category II: Barely tolerable consequences	Supporting elements Structural prefabricated products
Category III: Unacceptable or almost unacceptable consequences	Special buildings (nuclear power plants, etc.) Dams Tunnels Exceptional bridges and viaducts Monuments or prestigious buildings Railway sleepers

Table 3.4. *ISR: category of consequences*

Three additional classes relating to the exposure of concrete to the risks of ISR have been created; these classes (XH1, XH2, XH3) take into account the fact that one of the triggers is humidity.

These classes are defined as follows (Table 3.5).

Exposure classes	Description of the environment	Examples illustrating the choice of exposure classes
XH1	Dry or moderate humidity	– Part of structure inside buildings where the humidity level of the ambient air is low or medium – Part of structure located outside and sheltered from the rain
XH2	Alternating humidity-drying, high humidity	– Part of structure inside buildings where the humidity level of the ambient air is high – Part of structure not protected by coverage and subjected to the weather, without stagnation of water on the surface – Part of structure not protected by coverage and subjected to frequent condensation
XH3	In constant contact with water: – permanent immersion – stagnation of water on the surface – tidal zone	– Part of structure permanently submerged in water – Elements of marine structures – A large number of foundations – Part of structure regularly exposed to splash water

Table 3.5. *Exposure classes*

Each exposure class corresponds to a level of prevention for which the choice remains the responsibility of the developer.

Structure category \ Exposure class	XH1	XH2	XH3
I Low or acceptable risk	As	As	As
II Barely acceptable risk	As	Bs	Cs
III Unacceptable risk	As	Cs	Ds

Figure 3.29. *Level of prevention*

Each level of prevention is defined in the Guide in Table 3.6.

Level of prevention	Conditions
As	$T_{max} < 85$ °C or 85 °C < T_{max} < 90 °C and the time during which the temperature exceeds 85 °C is less than 4 h.
Bs	$T_{max} < 75$ °C or 75 °C < T_{max} < 85 °C and – Either the time during which the temperature exceeds 75 °C is less than 4 h and the active Na_2O level is less than 3 kg/m^3; – or sulfate resistant cement is used; – or cement type CEM II/B-V; CEM II B-S; CEM II B-Q; CEM II/B-M; CEM III/A or CEM V is used with SO_3 of cement less than 3% and C_3A less than 8%; – or the durability of the concrete is verified (performance test with respect to the ISR); – or combinations are used with CEMI of fly ash, slag and pozzolan such that the addition ratio is greater than 20% and C_3A is less than 8% and SO_3 is less than 3%.

Cs	$T_{max} < 70$ °C or 70 °C $< T_{max} <$ 80 °C and the same conditions as for level Bs.
Ds	$T_{max} < 70$ °C or 65 °C $< T_{max} <$ 75 °C and use of sulfate-resistant cement with an active Na_2O content of less than 3 kg/m^3. Validation of the formula by an ISR expert laboratory.

Table 3.6. *Conditions to respect level of prevention*

Figure 3.30. *Example: sulfate attack at the foot of the lifting screws of a waste water plant*

3.1.2.1.3. Alkali reaction

The mechanisms of alkali reaction are as follows:

– natural aggregates have reached a chemical equilibrium during their geological evolution. The latter is suddenly modified when it is incorporated into a strong alkaline cement matrix;

– the search for a new equilibrium involves reactions at the interface between cement and the aggregate, which can be beneficial (for example carboaluminate formation) or harmful (alkali reaction);

– this phenomenon is a liquid solid reaction between reactive silica forms of aggregates and the alkaline solution of the cement matrix. This results in the formation of calco-alkaline gels that can expand inside the concrete and lead to cracking.

The different types of alkali reactions are then as follows:

– silica alkali reaction: this occurs with rocks comprising amorphous silica forms such as opal, cristobalite and trydimite.

Mechanism: $SiOH + OH - SiO^- + H_2O$

$SiO^- + Na^+ - SiONa$

Simultaneously: $SiOSi + 2OH - SiO^- + \text{-}OSi + H_2O$

This double equation leads to the formation of a polymerized calco-alkaline gel;

– alkali silicate reaction: this involves metamorphic, sedimentary or igneous rocks;

– alkali carbonate reaction: this occurs with dolomitic rocks.

The different types of reactive silica are contained in the minerals presented in Table 3.7 (from the LCPC Recommendations).

NOTE.– Expansion properties of gels.

The swelling pressures that are induced by the alkali reaction and determined by theoretical calculation can vary from 45–140 MPa, whereas the experiment gives values of 3–10 MPa. The morphology of gels is also different from one concrete to another. Nevertheless, the pathology occurs due to the expansive nature of gels.

Rocks		Sensitive minerals in alkaline environment
MAGMATIC	Granite Granodiorites	Quartz with deformed network Altered fedspathic minerals, open grain joints
	Rhyolites Dacites Andesites Trachyandesites Basalts	Presence of siliceous glasses or more or less devitrified balsatic glasses, presence of tridymite, cristobalite, opal
	Obsidian Volcanic tuffs Pitchstone	Glasses rich in silica that are more or less devitrified and often microcracked
METAMORPHIC	Gneiss Mica-schists	Quartz Second-generation microquartz; open grain joints, fedspathic minerals and altered microacies
	Quartzites Hornstone	Quartz cement combined with quartz and opaline cement. Presence of second-generation microquartz. Presence of phyllosilicates. Presence of quartz or microcracked quartz
SEDIMENTARY	Sandstone Quartzites	Poorly crystallized siliceous cement, expanded grain boundaries
	Grauwackes Siltites Quartz schists	Associated mineral phyllites. Presence of opal, quartz microcrystallisins
	Chert Flint	Presence of chalcedony, opal
	Limestone Dolomitic limestones Dolomites	Presence of opal-type silica in micronodules or diffused throughout the network, associated or not with sedimentary sulfates and phyllites

Table 3.7. *Sensitive minerals table*

Criteria for quantifying aggregate reactivity have been established and are summarized in Figure 3.31.

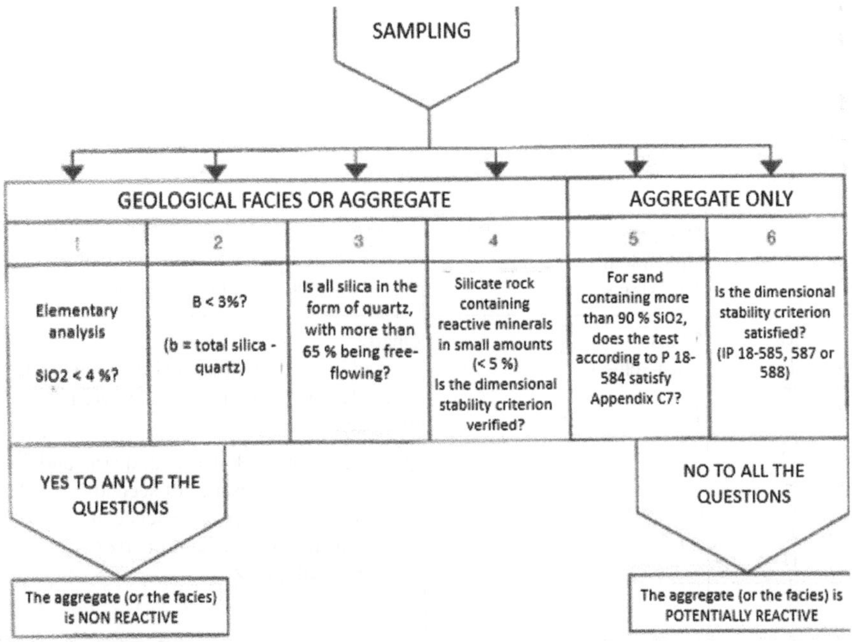

Figure 3.31. *1991 LCPC Recommendations.*
For new works, refer to FD P18-542 "Alkali Reaction"

The typology of damages is as follows:

– cracking in the form of macrocracking using mesh of several decimeters in size with traces of humidity and exudation of expansive gels;

– stress-oriented cracking for prestressed or heavily reinforced structures;

– possible formation of a burst cone for the aggregates close to the cladding (photo from LCPC).

NOTE.– Practical observations of alkali reaction gels under SEM.

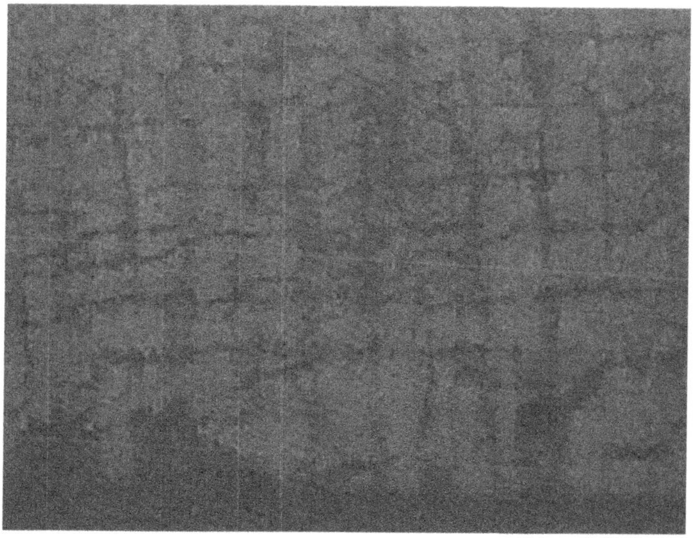

Figure 3.32. *Concrete cracking (St Hyacinthe-Quebec retaining wall)*

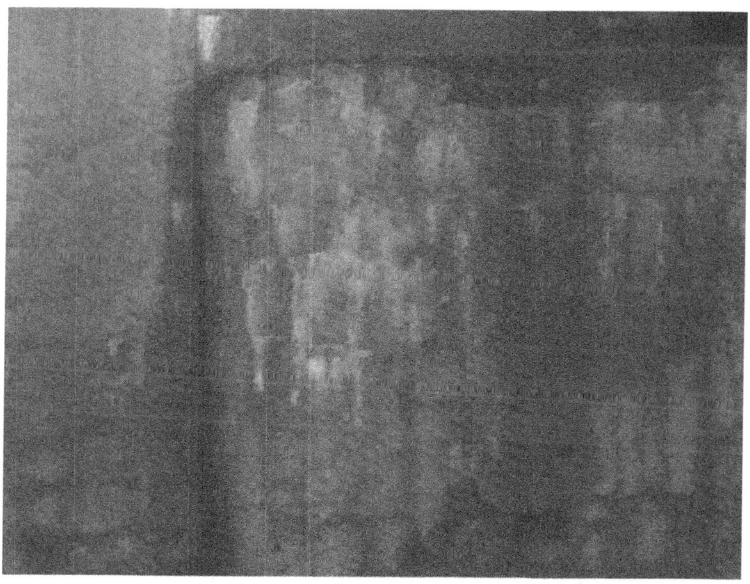

Figure 3.33. *Alkali reaction in a retaining wall (St Hyacinthe-Quebec)*

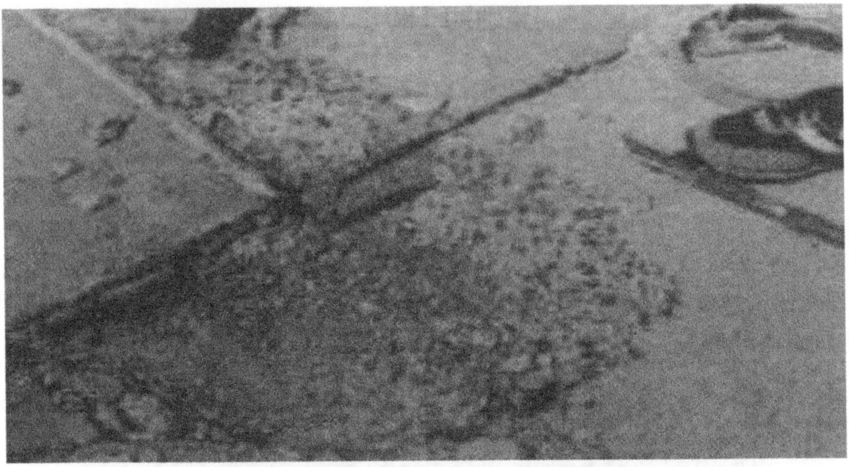

Figure 3.34. *Burst cones for aggregates*

The following photos are taken from observations under a scanning electron microscope coupled with chemical analyzes on polished sections and fresh fractures of a concrete sample.

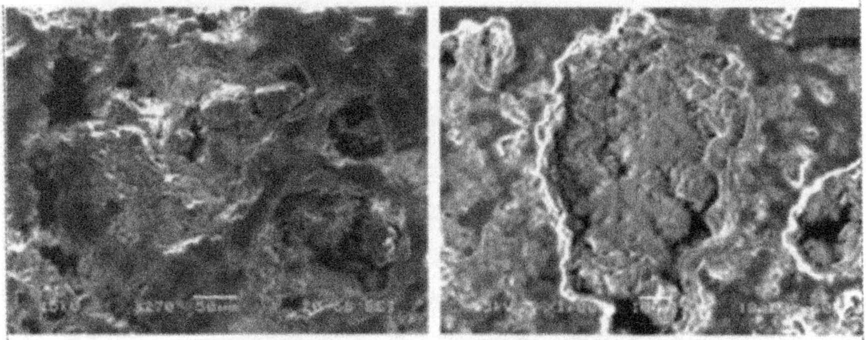

Photo 3: Alkali-reaction gel in aggregate
SEM, polished section, LV, LSEI, x270

Photo 4: Silico-calco-alkaline rosettes in aggregate
SEM, fracture, HV, LSEI, x1000

Figure 3.35. *Alkali-reaction from electron microscope*

The chemical analyses give the results shown in Figure 3.36.

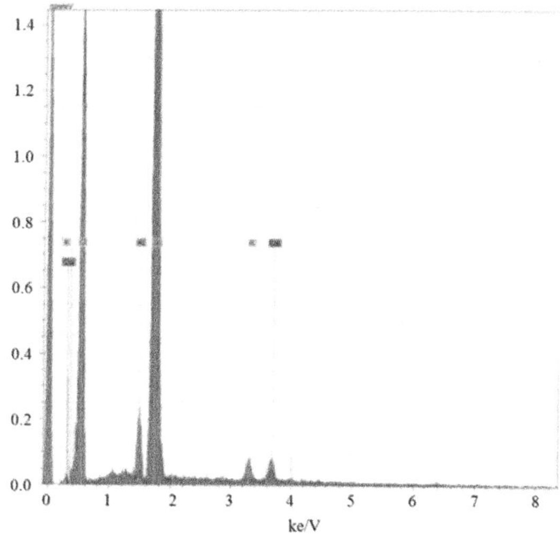

(a) EDS chemical analysis of aggregate (blue curve) and the alkali-reaction gel (brown line) of Photo 3

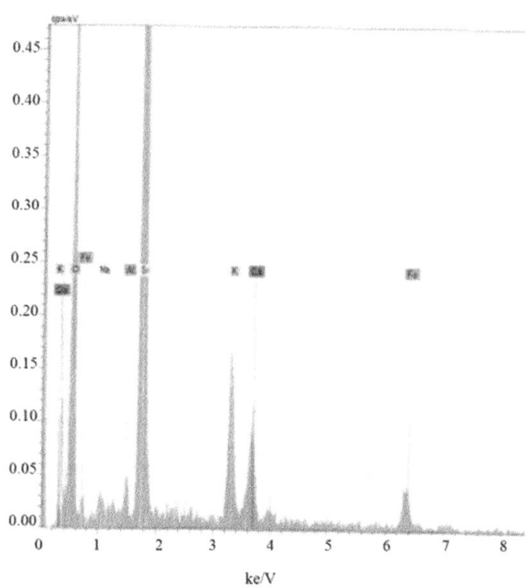

(b) EDS chemical analysis of silico-calco-alkaline rosettes from Photo 4

Figure 3.36. *Alkali-reaction from chemical analysis*

The previous elements show the presence of alkali reaction gels and silico–calco–alkaline rosettes in the concrete samples.

3.1.2.1.4. Attack from pure water and seawater

Because of its rich composition of sulfates and chlorides, seawater is one of the most aggressive environments for concrete.

A chloride attack is added to ettringite formation (seen previously) according to the process described next in the following.

Chlorides that are not chemically fixed in the cement matrix can migrate more or less deeply into the concrete by capillarity due to alternating moistening and drying.

They can then reach sufficient numbers to depassivate the reinforcements.

The capacity for fixing chloride ions is a function of the quantity of C_3A tricalcium aluminate that is present in the cement.

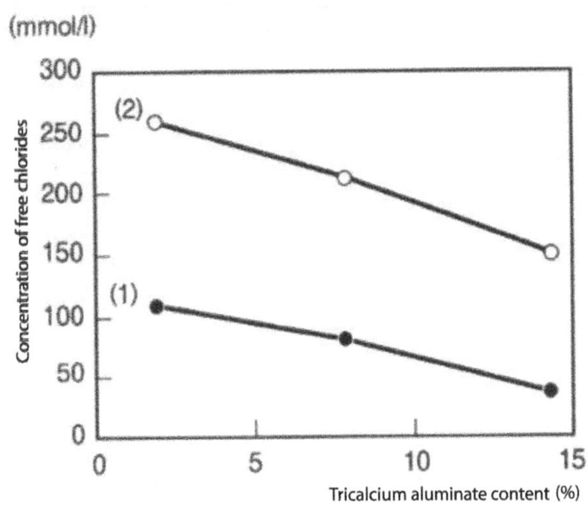

Figure 3.37. *Concentration of free chlorides in function of the quantity of C3A*

Moreover, the diffusion of the Cl^- ions in the cement matrix is directly related to the W/C ratio, as shown in Figure 3.38 (according to the LCPC document).

The setting time depends on the diffusion coefficient of free chlorides and, consequently, the porosity of the concrete.

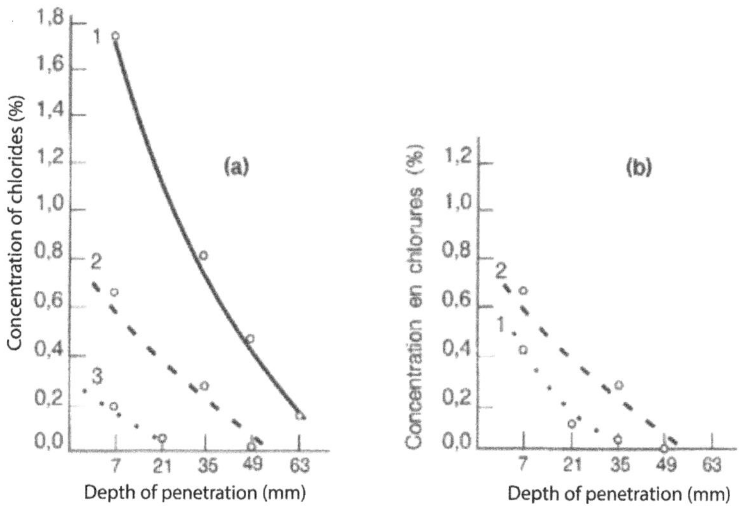

Figure 3.38. *a) Solution containing 150 g/L Cl⁻ with W/C ratios of 0.71, 0.47 and 0.23. b) Solution at 30 and 150 g/L with a W/C value of 0.47*

By analogy with the aforementioned carbonation phenomena, we find a similar curve for the duration of setting as a function of the cover.

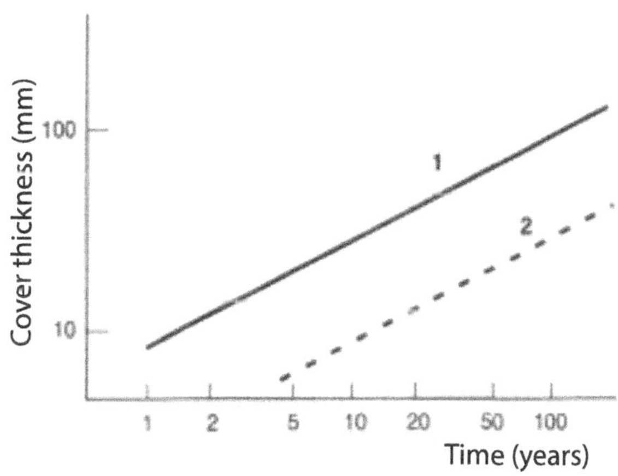

Figure 3.39. *Influence of cover thickness on the life span of a structure*

The seawater also contains sulfates that can generate sulfate reactions, as shown in Figure 3.40.

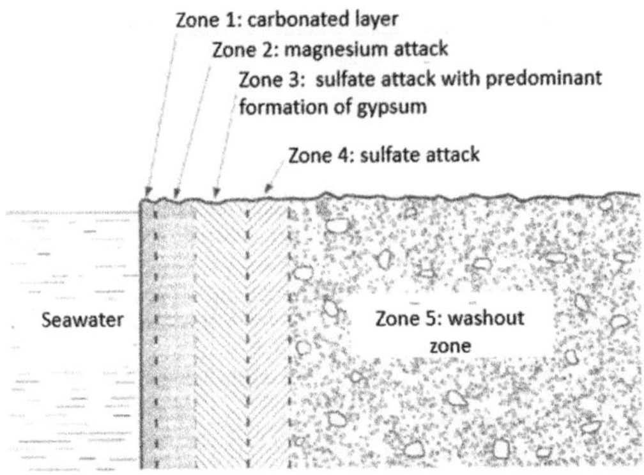

Figure 3.40. *Concrete attacks by sea water*

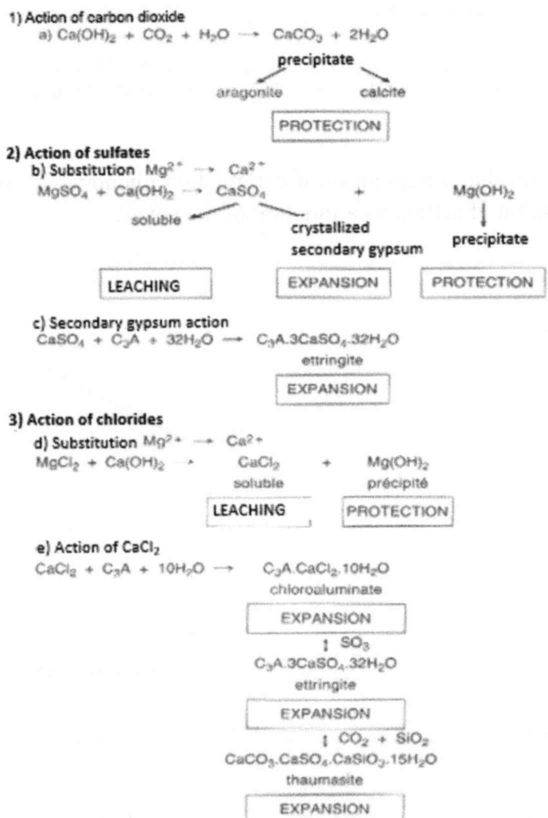

Figure 3.41. *Diagram of the deterioration of concrete by sea water*

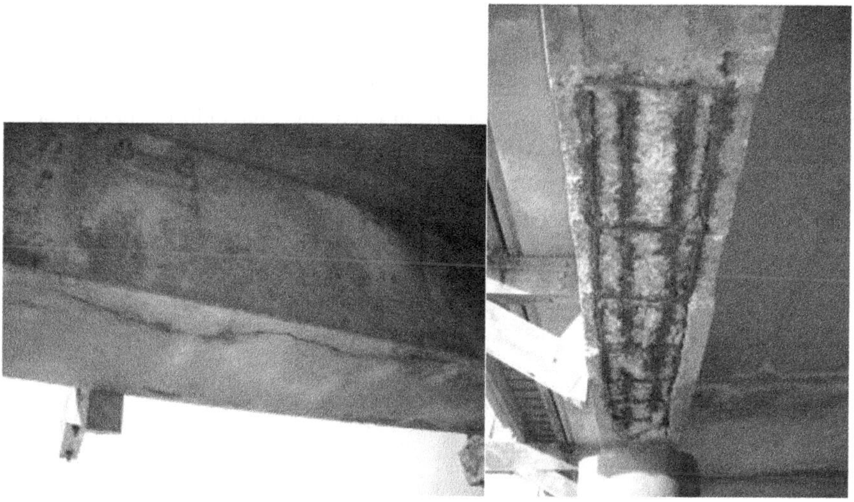

Figure 3.42. *Example of marine salt attacks*

Concrete samples can be taken by coring and the details presented in Tables 3.8 and 3.9 can be analyzed (Eurofins example).

NOTE.– Interpretation of results.

The above results allow us to highlight the following elements:

– the composition of concrete comprises a mixture of cement and siliceous aggregates (about 94% of silicate elements). The cement dosage is relatively low, around 300 kg/m^3 for a water content of 242 L/m^3, so a W/C ratio of around 0.80. The EN 206.1 standard for concrete would classify this environment in *XS3* (tidal zone). This classification leads to an equivalent binder dosage of 350 kg/m^3 for a W/C ratio of 0.50 and a minimum compressive strength of 35 MPa. As a result, the concrete is largely underdosed for cement and overdosed for water, resulting in a concrete that is not very compact and is very sensitive to seawater;

– determining the hydration rate gives a value of about 18% for an expected value of 17%, which validates normal setting of the binder in the poured concrete;

– the porosity of the concrete is 15.4% for a density of 2,187 kg/m^3. These values corroborate the previous remark on the production of a relatively porous concrete;

– the determination of free chlorides (soluble in seawater) was carried out at three different depths: at the surface (sea side), at –5 cm and at –10 cm from the surface. The values obtained are significant since they are 1.14% at the surface, 0.52% at –5 cm and 0.48% at –10 cm. The average chloride dosage across the entire core sampling is about 0.63%. For reinforcements located 10 cm below the surface, the chloride content relative to the cement dosage is 3.5%. Corrosion of the reinforcements by chlorides is observed for the sample and it should be noted that it is at –10 cm from the surface (the regulatory cover for a class XS3 concrete is 5 cm).

Page 1 of 2

CONCRETE ANALYSIS REPORT®

Date received: 08/06/2009 Site of sampling: NR

File reference: Andernos-les-Bains jetty – tidal zone Sample reference: Coring on 2 June 2009

Date of sampling: NR Concrete age: NR

Sampling done by: NR

WEIGHT COMPOSITION OF CONCRETE		
Parameters	Results	Units
Cement content	*13.8*	%
Aggregates content	*75.2*	%
silicate aggregates	*94.2*	%
calcite aggregates	*5.8*	%
Water content	*11.1*	%
Total	*100*	%

CONSTITUTIONAL PARAMETERS		
Parameters	Results	Units
Cement content	*301*	*kg/m³*
Water content	242	*L*
W/C ratio	*0.80*	--
Hydration rate	18	%
Entrained air content estimate	*8.4*	%

PHYSICAL PARAMETERS		
Parameters	Results	Units
Open porosity (6)	*15.4*	%
Visible density (6)	*2187*	kg/m³
Soaked density (6)	2341	kg/m³
Elution (length/diameter)	2	--
Compressive strength (7)	*NR*	*N/mm²*
Resistance class	*NR*	--

Table 3.8. *Concrete chemical analysis*

LEM file number: 09S019417-001

Version: 15/06/2009 V1

CHEMICAL PARAMETERS		
Parameters	Results	Units
Basic parameters Fire loss 105°C/ 450°C (1)	2.42	%
Fire loss 450°C/550°C (1)	0.28	%
Fire loss 550°C/960°C (1)	2.09	%
Soluble silicate content (4)	3.91	%
Na^2O content (3)	2.66	%
MgO content (3)	1.57	%
Al_2O_3 content (3)	6.90	%
SiO_2 (insoluble) content (3)	47.2	%
P_2O_5 content (3)	0.213	%
SO_3 content (3)	2.18	%
K_20 content (3)	1.36	%
CaO content (3)	25.4	%
TiO_2 content (3)	0.221	%
MnO content (3)	0.067	%
Fe_2O_3 content (3)	1.58	%
ZnO content	0.022	%
SrO content (3)	0.022	%
PbO content (3)	0.01	%
Total	98.11	%
Complementary parameters *Insoluble residue* (2)	79.1	%
Fire loss (1)	5.04	%
Bound water content (1)	2.42	%
Organic content (1)	0.28	%
CO_2 content (1)	2.09	%
$CaCO_3$ equivalent content (1)	4.74	%

PATHOLOGICAL PARAMETERS		
Parameters	Results	Units
Carbonation depth (5)	0	mm
Sulfate content (2)	0.56	%
Free chlorides content (average) (2)	0.63	%
Chloride/cement content	4.58	%
Chloride class	\	--
Soluble Na_2O content (2)	0.10	%
Soluble K_20 content (2)	0.06	%
Soluble Na_20 equivalent content (2)	0.19	%
Soluble Na_20 equivalent content	4.16	kg/m^3

NR: Not reported Retention of samples: 1 month after the date of analysis report

CODE	1	2	3	4	5	6	7
Internal method adapted from	Bull LPC no. 5, Jan-Feb 1984	NF EN 196-2	P 15-467	NF T90-007	NF EN 14630	AFREM recommendations 1997	NF EN 12390-3

COMPLEMENTARY PARAMETERS- free chlorides at 3 depths		
Sampling area	Results	Units
Surface	1.14	%
Middle: 5 cm	0.52	%
Reinforcement: 10 cm	0.48	%

Table 3.9. *Concrete chemical parameters*

3.1.2.1.5. Freeze-thaw cycle attack

These are degradations caused to concrete through the penetration of surface water into the concrete through the capillary network and the cracks. This causes stresses by swelling.

The most common symptoms are:

– flaking of the concrete surface;

– swelling of the structure followed by network cracking.

The parameters that influence the freezing mechanisms are as follows:

– porosity of the cement matrix and more particularly, the distribution and size of pores (critical spacing of air bubbles);

– the degree of critical saturation or the ratio between the amount of frozen water and the volume available for expansion;

– transformation of water into ice as a function of temperature, pressure, pore size;

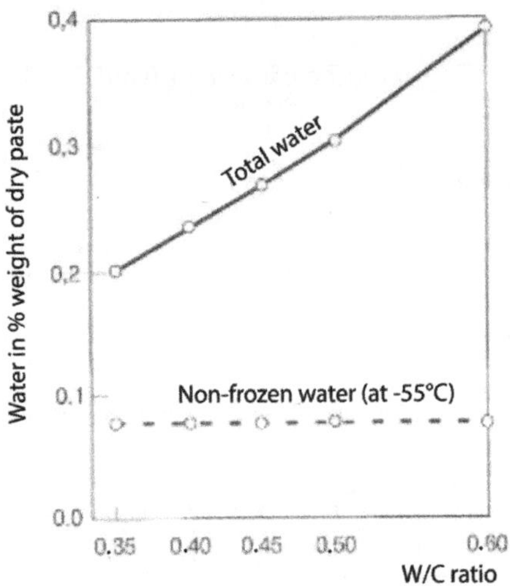

Figure 3.43. *Total water and non-frozen water. The fraction of non-freezeable water (here 8%) can reach 20% in a fully hydrated cement paste*

There is a decrease in the melting temperature of ice with decreasing radius of pores (Figure 3.44).

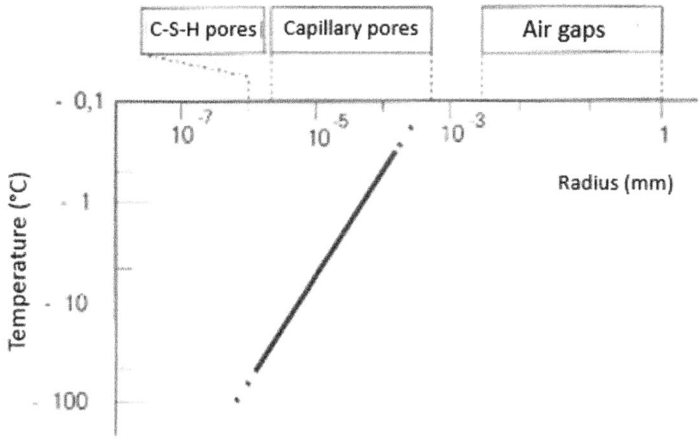

Figure 3.44. *Influence of the radius of pores on the melting temperature*

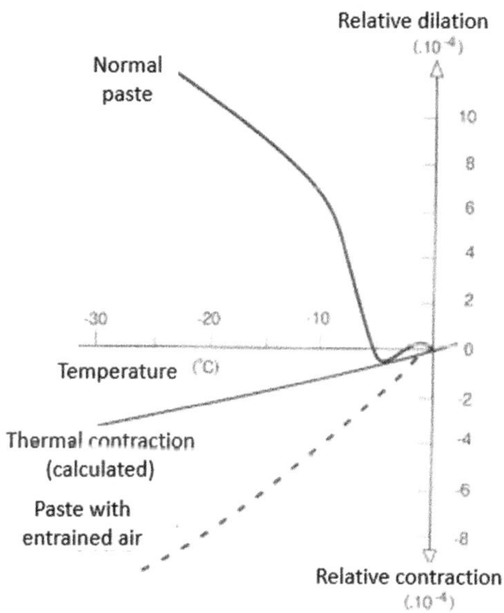

Figure 3.45. *Contraction/dilatation in function of temperature*

– there is transformation of water into ice with an expansion of 9% by volume and expulsion of water from the capillaries with the relevant mechanical constraints. We note here the influence of entrained air;

– influence of the cooling rate;

– the number of freeze-thaw cycles.

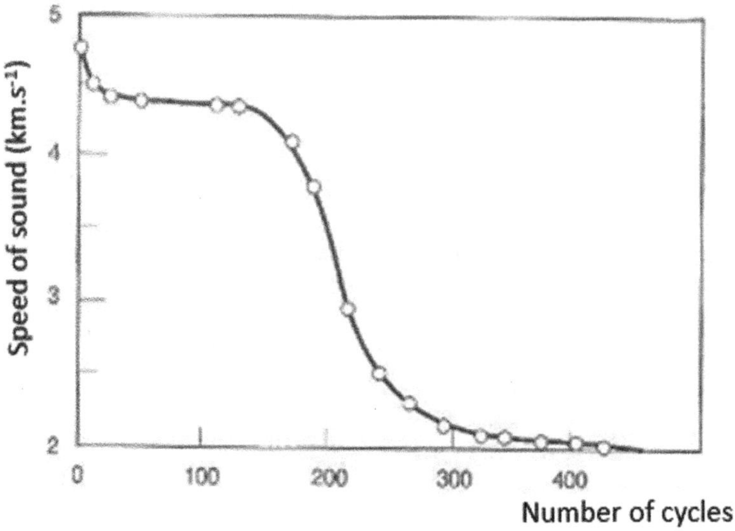

Figure 3.46. *Influence of the number of freeze-thaw cycles on the speed of sound in concrete structures*

3.1.2.2. *Degradation by mechanical aggression*

These are mainly abrasion and erosion phenomena for structures that are in contact with intense water circulations, possibly laden with sandy particles.

The pathology appears in the form of surface wear, flaking of the concrete.

Another mechanical attack is shock: for example, the accidental shock of a truck on the cask of a water tower or on the columns of a skip building.

3.1.2.3. *Bacteriological attacks*

Although these are infrequent, bacteriological attacks have been seen in concrete structures that are in contact with urban waste water.

This type of attack is caused by *Thio Bacillus* that, by oxidation of H_2S to H_2SO_4 on the condensing walls, attacks the Portlandite of the concrete to form ettringite (see above).

3.2. The pathology of masonry structures

3.2.1. *General information*

The use of masonry in water treatment or pumping stations is generally limited to structures connected to the reservoirs because of its inability to ensure reservoir-specific sealing stress. It allows technical rooms to be made (workshops, laboratory, local pumps, etc.).

3.2.2. *Major disorders that may affect masonry*

3.2.2.1. *Cracking*

The aforementioned principles for reinforced concrete are also applicable to masonry walls.

In addition, it should be noted that the whole construction is no longer monolithic but is the result of an assembly of manufactured materials (bricks, concrete agglomerates, etc.) and assembly joints. The assembly is generally coated for rainwater waterproofing.

Let us discuss the different types of cracking according to their layout (see definition above):

3.2.2.1.1. Multidirectional cracks

These elements thus have no preferred direction. They are mainly found in coatings. They are cracks in the shrinkage of coatings based on hydraulic binders that can occur for the following reasons:

- overdosage of last layer of cement;
- evaporation is too fast;
- insufficiently wet support;
- sand is too thin;
- layer is too thin.

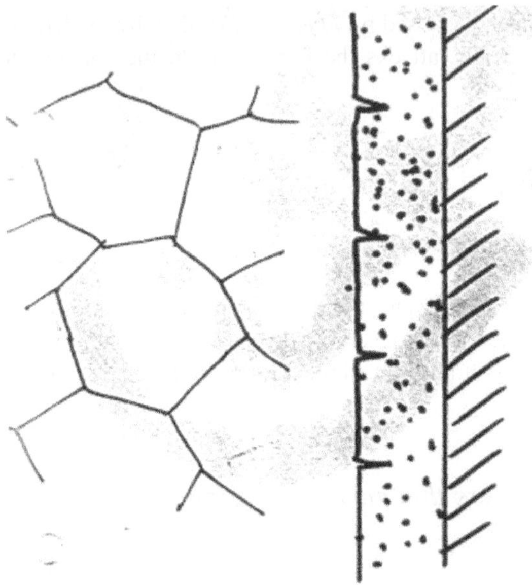

Figure 3.47. *Example of "faience" of the plaster*

Figure 3.48. *"Faïence" photo of the plaster*

3.2.2.1.2. Horizontal cracks

These cracks are divided as follows:

– cracks that result from simple heterogeneity of the support: they are found in the coating at junctions of materials with different thermohygrometric behaviors. For example, a junction between a concrete lintel with a masonry wall located above it;

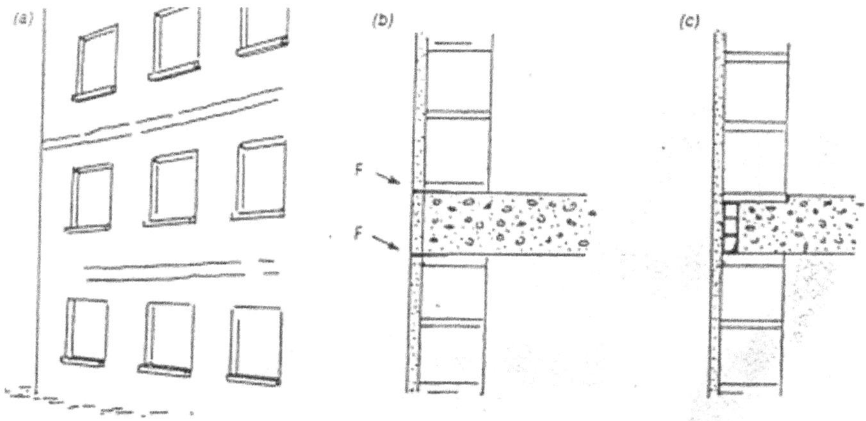

Figure 3.49. *Horizontal cracks associated with the juxtaposition of different materials under the coating*

Figure 3.50. *Cracking of the coating at the connection between reinforced concrete structures and masonry filling*

– cracks resulting from the interaction between wall and floor: these emerged recently when modern constructions favored walls that were increasingly thin and with increasingly "nervous" floors. The difference in inertia does not allow the wall to avoid floor bearing rotation. This creates horizontal cracks under the slab support approximately two rows below;

– cracks that result from a production defect: resetting after stopping the assembly on a same foundation;

– cracks associated with foundation problems on swelling soil: these tend to appear above the libations or at the lintels and mainly occur due to a rotation of the foundations.

Figure 3.51. *Cracking connected to the foundation mode on swelling soils*

3.2.2.1.3. Vertical cracks

These cracks are also divided as follows:

– cracks resulting from heterogeneity of the support (ditto horizontal cracks);

– cracks that occur due to thermal phenomena: absence of expansion joint on an exposed facade, for example;

– cracks linked to the saber shot: precut masonry elements applied in line with a joint;

– cracks linked to insufficiently dry masonry (Figure 3.53).

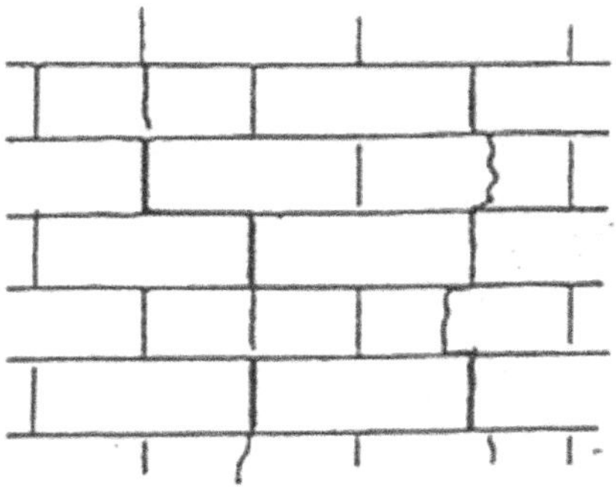

Figure 3.52. *Example of a saber shot*

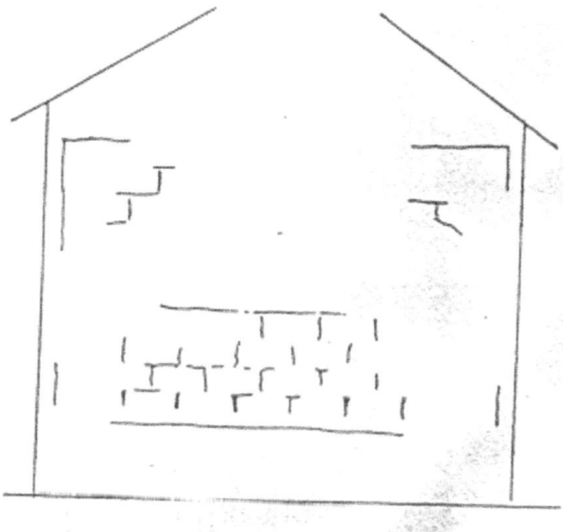

Figure 3.53. *Cracking of coating in masonry blocks that are insufficiently dry*

3.2.2.1.4. Slanted cracks

There are two main causes for these cracks:

– shrinkage: particularly at the bay angles (shrinkage due to the casted support piece). These are then inclined at 45° to the outside of the bay and follow the masonry joints;

– shear stresses that occur due to different loads on the same masonry panel (for example lowering of a load over a trumeau and under a spandrel) but also due to differential settlement of foundations (see above for concrete).

Figure 3.54. *Differential settlement cracking at building angle*

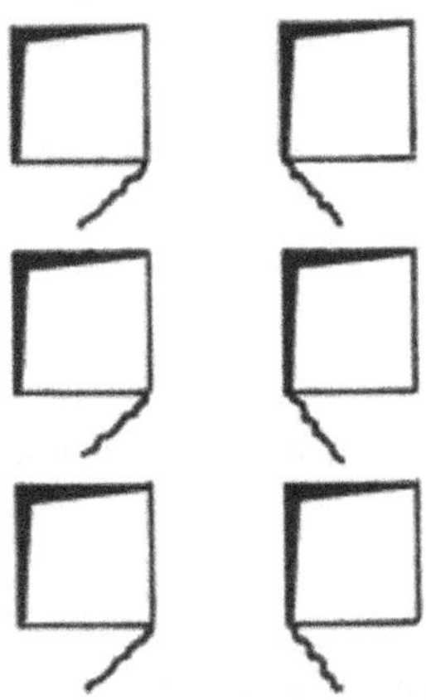

Figure 3.55. *Cracks at 45° under different loads*

3.2.2.2. *Waterproofing*

These mainly involve:

– waterproofing defect on the facades due to an error in the design of the wall (thickness, exposure, etc.);

– capillary uplift through the masonry.

However, these elements will not be discussed here because of their low involvement in wastewater plants and civil engineering works.

Figure 3.56. *Lack of waterproofing in a buried construction*

3.3. The pathology of composite material structures

3.3.1. General information on composite materials

Pathologies that are frequently observed on concrete and metallic structures have led to the development of composite materials in the building mode and in civil engineering.

By definition, we consider a composite material as any material consisting of two elements for which, once composed, the properties are superior to those of the elements alone.

These materials, which are mostly from the aeronautical world (such as Freyssinet's carbon fiber fabric, for example), are used in the construction world in several forms:

– reinforcement of existing structures (wood, concrete, metal);

– full structure from reused concrete, metal or wood structures;

– covers, manholes, antacid protection, on heavily attacked structures.

3.3.1.1. *Composition of composite materials*

Composite materials, in general, are formed from:

– a matrix;

– reinforcing fibers.

3.3.1.1.1. The matrix

The role of the matrix is to surround and protect the fibers.

It also allows:

– transfer of stresses between fibers;

– protection of fibers against ambient conditions;

– mechanical protection of fibers (shocks, etc.);

– prevention of the deformation of fibers.

The choice of matrix must be made according to the end use of the final product.

There are a large number of matrices that can be classified into four major families:

– Polymer matrices:
- polyester
- vinylester
  - epoxy
  - phenolic
  - polyamide

  (polyester, vinylester, epoxy, phenolic, polyamide: thermostable)
  - thermoplastic (polycarbonate)
  - polyurethane
  - silicon

– Metallic materials:

- aluminum;
- titanium;
- magnesium;
- stainless steel.

– Ceramic matrices:

- alumina (Al_2O_3);
- silica carbide (SiC);
- silica nitrate (Si_3Ni).

– Mineral materials:

- cement mortar;
- clay mortar.

In civil engineering, the most widely used matrices are those in groups 1 and 4. We will not discuss those in group 4 here since their applications mainly involve mortar screeds reinforced with metallic or synthetic fibers or baked clay elements.

For the first group, let us establish a quick comparison between thermostable matrices and thermoplastic matrices.

Thermostable materials	Thermoplastic materials
Very low viscosity before drying	Very short drying time
Thermal stability	Low drying shrinkage
Good chemical resistance	Good resistance to chlorine
Good impregnation capacity	Good deformation capacity
Easy to manufacture	Unlimited storage time
Economic	Ability to modify shapes
	Possibility of repair and recycling
	Good resistance to tearing

Table 3.10. *Different types of matrices*

As a result, to date, the most widely used matrices in the world of construction are thermostable resins (cost). Among the most common ones, each has their own advantages, which should be taken into account to ensure compatibility with the work to be carried out.

Polyester resins	Vinylester resins	Epoxy resins	Phenolic resins
– Low viscosity (malleable) – Fast drying time – Significant shrinkage upon drying – Possibility of drying at ambient temperature or at high temperatures – Good electrical resistance – Good resistance to fire – Good value for money – Good mechanical properties, although inferior to other resins	– Good mechanical properties – Excellent fluidity – Good bonding with reinforcing fibers – Good resistance to corrosion – Good resistance to chemical agents – Fast drying time – Significant shrinkage upon drying	– Excellent mechanical properties (resistance, etc.) – Little shrinkage during drying – Good electrical behavior – Long drying times – High cost – Good bonding with all fibers – Good resistance to chemical agents and solvents – Low resistance to high temperatures	– Excellent electrical characteristics – Good resistance to high temperatures – Good resistance to abrasion – Good resistance to chemical agents and organic solvents – Excellent bonding with other resins – Dimensional stability – Weak mechanical properties

Table 3.11. *Characteristics of different resins*

	Polyester	Epoxy	Phenolphthalein	Vinylester	Polyamide
Resistant to	Water Fuel Petrol	Alcohol Petrol Benzol Mineral water Grease	Water Oil Grease Petrol Benzol Alcohol	Seawater Tar Chlorine dioxide	Ether Alcohol Kerosene
Not resistant to	Acids Bleach Benzol Alcohol Toluene	Ester Alkaline compounds Oils	Concentrated alkaline acids and compounds	H_2SO_4 (75 %) NaOCl (6 %) NaOH (15 %)	Alkaline compounds Ammonia Humidity

Table 3.12. *Chemical resistance of different resins*

3.3.1.1.2. Fibers

Fibers are incorporated into the matrix to reinforce it. The fibers may or may not have a preferential orientation. The influence of the direction of fibers on the mechanical properties of the material must be taken into account.

The main reinforcing fibers used in composite materials are:

– fiberglass (A, E, B, S, R type fibers);

– synthetic fibers (polypropylene, polyethylene, nylon, polyester);

– carbon fiber;

– aramid fibers.

Fiberglass is the most widely used type of fiber in composite materials for construction.

Fiberglass	Carbon fiber	Aramid fibers	Synthetic fibers
– Good mechanical resistance – Good density – Incombustible – Good dimensional stability – Imputrescible – Economical – Good resistance to chemical agents – Less resistance to fatigue than other fibers – High hardness	– Good rigidity – Good density – High mechanical resistance – Low expansion coefficient – Good dimensional stability – High cost – Low impact resistance – Promotes galvanic corrosion with metals	– Low weight – Good impact resistance – Good tensile strength – Very low compressive strength – Resistance to chemical agents – Mechanical stability between –30 and 200 °C	Polypropylene– – Inert material – Good mechanical properties Polyethylene– – In cement additive and in geotextile Nylon– – Good tensile strength (synthetic geo) Polyester– – Most used but not the most performant

Table 3.13. *Summary table of the advantages and disadvantages for each type of fiber*

3.3.2. *Main pathologies of composite materials*

3.3.2.1. *Chemical incompatibilities between matrices and fibers*

Any fiber can reinforce any matrix.

These compatibilities are summarized in Table 3.14.

The pathology that results from an incompatibility generally leads to a lack of bonding between the fiber and the matrix, as well as premature aging of the complex.

There can also be galvanic corrosion between aluminum and carbon fibers.

Fibers \ Matrices	Unsaturated polyester (UP)	Epoxy resin (EP)	Phenolic resins (PF)	Vinyl ester (VE)	Polyimide (PI)	Polypropylene (PP)	Polyamide (PA)	Polyether ether ketone (PEEK)	Polyether sulfone (PES)
Fiber glass C	X	X	X	G	X	X	X	X	X
Fiber glass E	G	G	G	G	X	G	G	P	P
Fiber glass S	P	G	P	P	X	X	G	X	G
Carbon fiber HT	G	G	G	G	G	P	G		G
Carbon fiber HST	P	G	G	G	X	X	G	X	G
Carbon fiber HM	P	G	G	P	X	X	G	X	G
Aramid	P	G	P	G	G/P	X	G	X	P
Polyethylene	P	G	X	P	X	X	X	X	X

Good compatibility = G; Possible compatibility = P

Table 3.14. *Compatibility between resins and fibers*

As a result, the good matrix fiber parities are presented in Table 3.15.

<table>
<tr><th colspan="2">Matrix</th><th>Fiber</th></tr>
<tr><td rowspan="7">Thermostable</td><td>Polyester</td><td>Glass fiber</td></tr>
<tr><td rowspan="2">Vinyl ester</td><td>Glass fiber</td></tr>
<tr><td>Aramid</td></tr>
<tr><td rowspan="3">Epoxy</td><td>Glass fiber</td></tr>
<tr><td>Aramid</td></tr>
<tr><td>Carbon</td></tr>
<tr><td>Phenolic</td><td>Glass fiber</td></tr>
<tr><td rowspan="3">Thermoplastic</td><td>PVC</td><td>Glass fiber</td></tr>
<tr><td>PP</td><td>Glass fiber</td></tr>
<tr><td>Nylon</td><td>Glass fiber</td></tr>
<tr><td colspan="2">Cemento</td><td>Glass fiber AR</td></tr>
</table>

Table 3.15. *Association matrix/fiber*

3.3.2.2. *Chemical incompatibilities between composite materials and the surrounding environment*

Most composites are insensitive to common chemicals.

However, certain products tend to irreversibly damage certain complexes.

For example, paint strippers will attack epoxy resins.

The main causes of damage are summarized in Table 3.16.

	Polyester	Epoxy	Phenolphthalein	Vinylester	Polyamide
Resistant to	Water Fuel Petrol	Alcohol Petrol Benzol Mineral water Grease	Water Oil Grease Petrol Benzol Alcohol	Sea water Tar Chlorine dioxide	Ether Alcohol Kerosene
Not resistant to	Acids Bleach Benzol Alcohol Toluene	Ester Alkaline compounds Oils	Concentrated alkaline acids and compounds	H_2SO_4 (75%) NaO Cl (6%) NaOH (15%)	Alkaline compounds Ammonia Humidity

Table 3.16. *Chemical compatibilities and incompatibilities*

3.3.2.3. *Modification of polymers by radiation*

The main chemical bonds that are present in composite materials have bonding energies close to 100 kcal/mol (that is a few electronvolts).

Using a source of energy that is greater than that of the chemical bonds is therefore capable of chemically modifying the polymer.

In nature, light has sufficient energy to attack the internal bonds of composites.

Radiation	Wavelength (nm)	Energy (eV)
Gamma	10^{-4}–10^{-2}	10^{5}–10^{8}
Electrons	10^{-3}–10^{-1}	10^{4}–10^{7}
UV	10–100	5–10^{-5}
Visible	400–750	1–5
Infrared	750–10^{5}	10–0.2
Microwaves	> 10^{5}	<10^{-2}

Table 3.17. *Energy from radiations*

As a result, in the absence of any additional protection on the first composite materials, a pathology was observed, which resulted in a loss of elasticity of the material and rapid delamination.

NOTE.– This is still the case, for example, on boat sails made up of exotic fibers that are not protected from solar radiation.

3.3.2.4. *Osmosis*

This phenomenon is mainly directed to fiberglass laminates, which are embedded in a polyester matrix that is protected by a gel-coat type of paint in the presence of water.

It is assumed that water migrates through the gel-coat in the form of vapor and attacks the laminate to form solutes. In that state, the solutes are trapped behind the gel-coat membrane.

This results in the appearance of blisters filled with acetic acid, which can ultimately cause irreversible deterioration of the laminate.

Figure 3.57. *Picture of osmosis blistering*

3.3.2.5. *Delamination*

The delamination of composite materials was the main topic of numerous studies.

Among other things, it involves identifying delamination criteria based on the interface shear or on the calculation of the rate of redistribution of energies at the interfaces between the fiber and the matrix.

A delamination test is defined in the NFT 57-104 standard. It appears as shown in Figure 3.58.

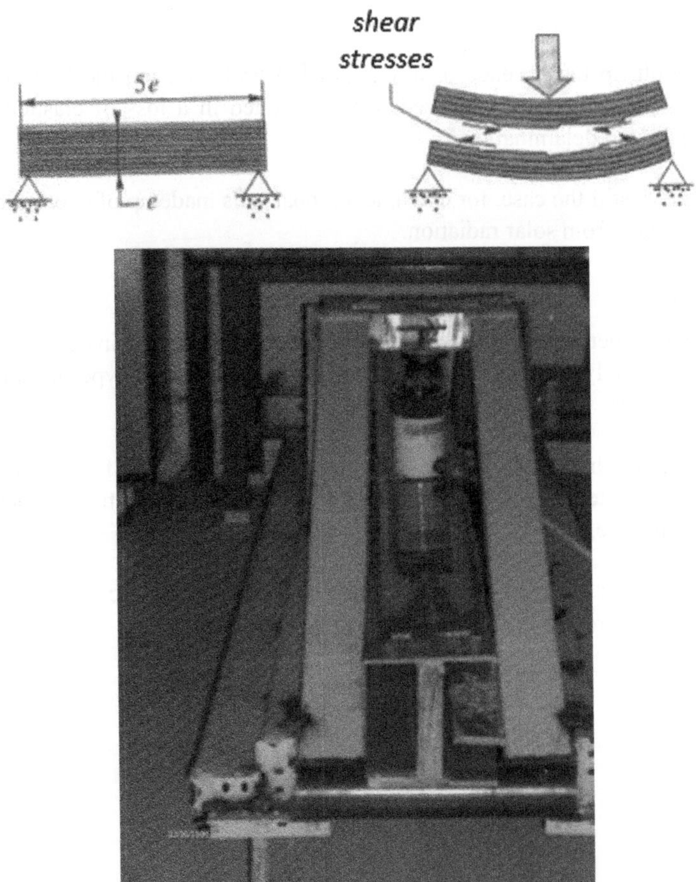

Figure 3.58. *Delamination test under the effect of a distribution of bending stresses*

From this, a conventional value for interlaminar rupture stress is deduced.

Delamination is the cause of many pathologies related to:

– stress imposed on the material that is greater than the calculated stress;

– a laminate manufacturing defect;

– use of the composite that does not correspond to its initial end-use.

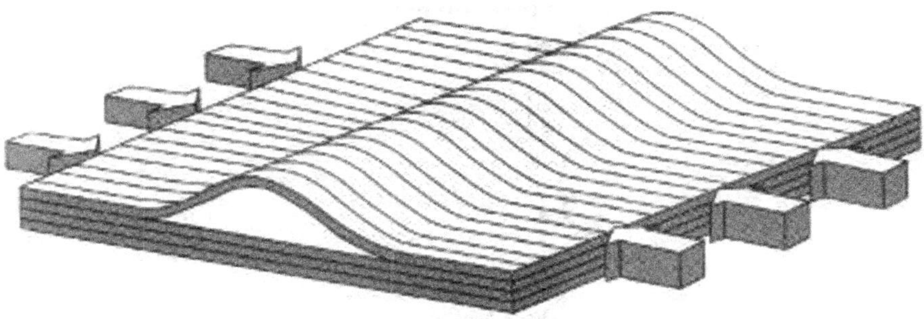

Figure 3.59. *Example of delamination by local buckling (composite materials document)*

3.3.2.6. *Breaking of the laminate*

Composite materials may have an isotropic or anisotropic behavior depending on the nature of the reinforcement inside the fold.

Indeed, the matrix reinforcing mixture can appear as:

– unidirectional reinforcement + matrix;

– fabric reinforcement (warp + weft) + matrix;

– capping reinforcement + weft.

A unidirectional fold will have a preferred direction of constraint, as shown in Table 3.18.

As a result, the direction and orientation of the stresses must be defined very accurately and the reinforcement of the composite should be positioned accordingly.

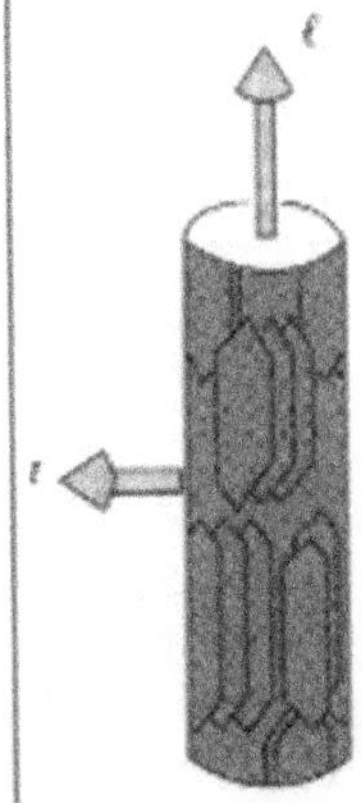

single fiber

	GLASS "E"	KEVLAR	"HR" CARBON	"HM" CARBON
longitudinal module of fiber in direction ℓ Ef_ℓ (Mpa)	74 000	130 000	230 000	390 000
longitudinal module of fiber in direction t Ef_t (Mpa)	74 000	5 400	15 000	6 000
shearing module of fiber $Gf_{\ell t}$ (Mpa)	30 000	12 000	50 000	20 000
Poisson coefficient of fiber $\nu f_{\ell t}$	0.25	0.4	0.3	0.35
	isotropic fiber	anisotropic fiber		

Table 3.18. *Mechanical characteristics of single fiber*

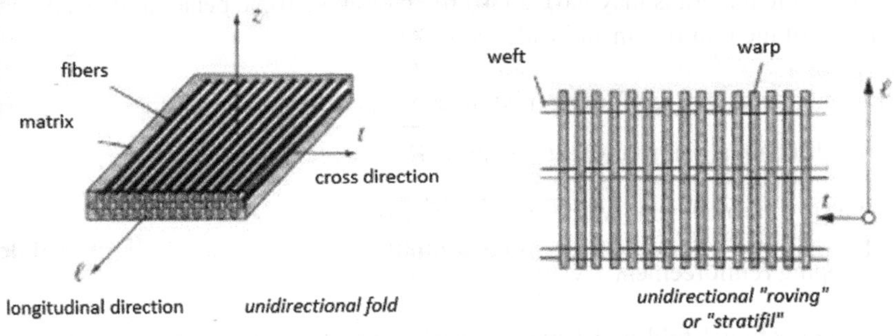

Figure 3.60. *Different types of repartitions*

EXAMPLE.– Pathology related to an error in positioning of the reinforcement (Figure 3.61).

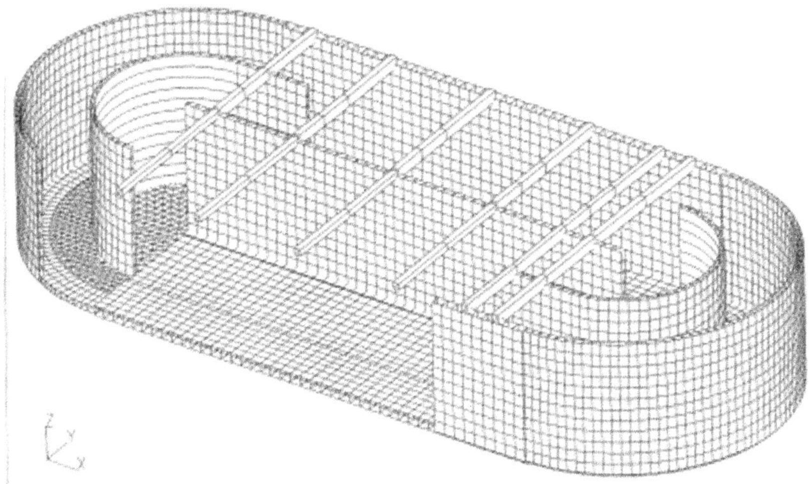

Figure 3.61. *Finite elements model of a tank*

Calculation of stresses.

– For a rectangular wall (Figure 3.62).

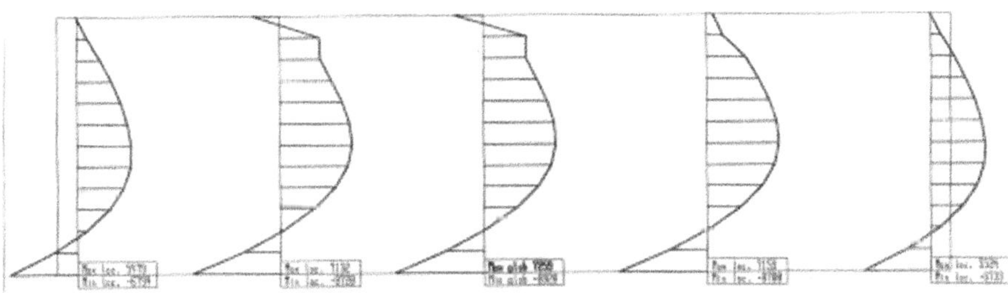

Figure 3.62. *Stress diagrams*

– For a semicircular area (Figure 3.63).

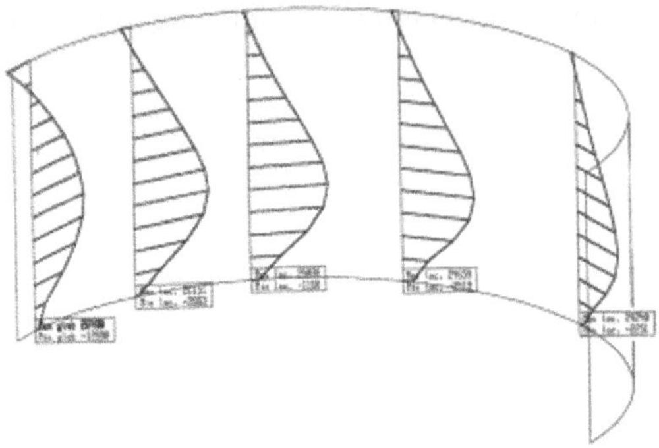

Figure 3.63. *Stress diagrams in the semi-circular part*

At the junction between the rectangular part of the tank and the semicircular part, there is a sudden change in stresses.

The absence of reinforcement of the fiber in this area has led to an opening of the structure (tearing of the composite) under the effect of the thrust from the water.

EXAMPLE.– A pathology linked to a lack of fiber continuity (Figure 3.64).

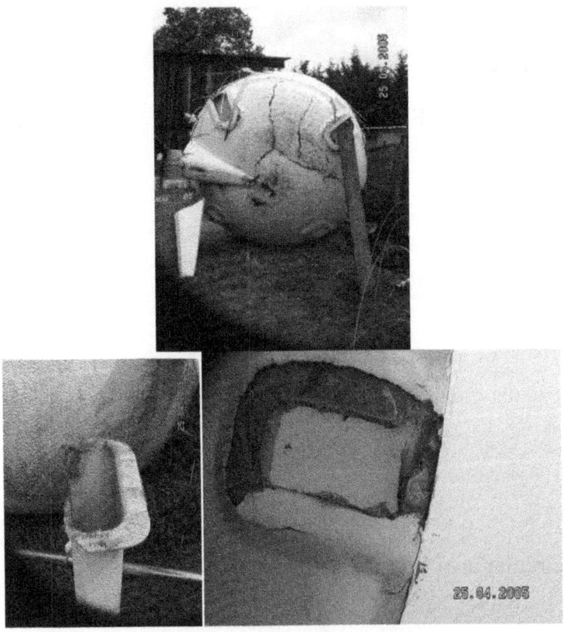

Figure 3.64. *Pathology linked to a defect in characterization of stresses at the foot and a defect in the use of tissues*

3.3.2.7. *Breaking of assemblies*

An assembly of composite materials remains a source of difficulty that frequently generates pathologies.

These assemblies can be carried out in several ways:

– by riveting or bolting metal parts;

– by collage.

In the first case, the drilling required for the assembly remains a factor of embrittlement of the material. The local resistance loss can be estimated at about 50% tensile and 15% compression.

As with traditional materials (steel, wood), the hole is the site of stress concentration, which causes cracking of the laminate as shown in Figure 3.65 (diametrical pressure and capping pressure).

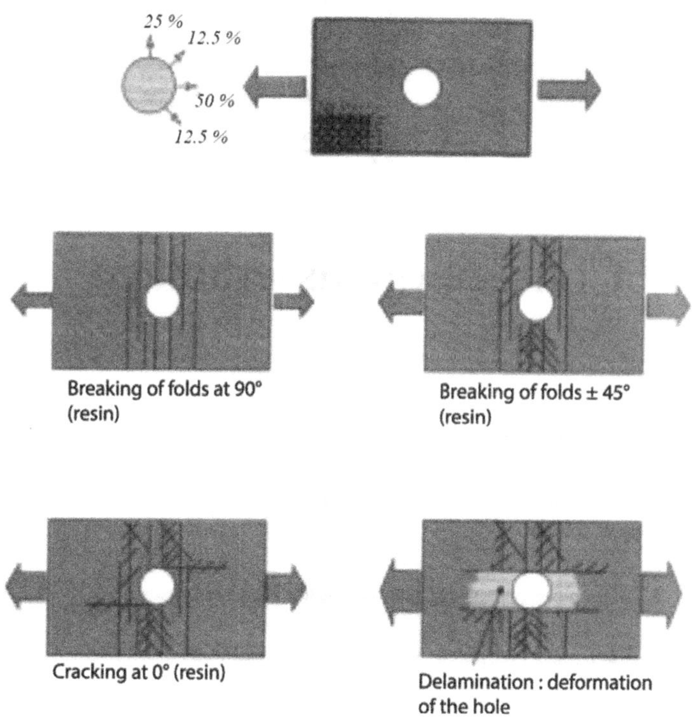

Figure 3.65. *Composite materials document*

As a result, the following modes of assembly breakage can occur.

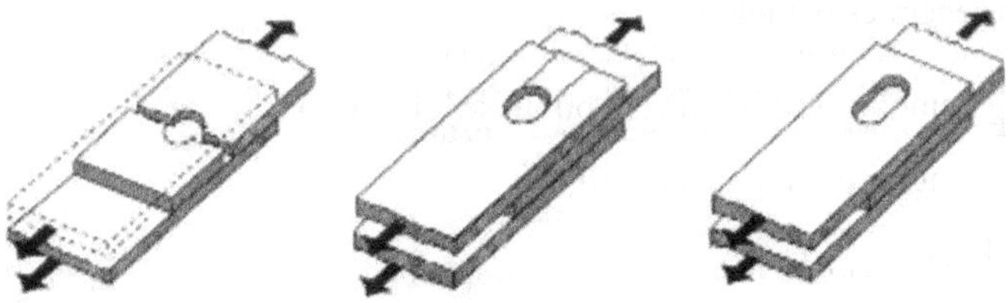

Figure 3.66. *Assembly breakage*

The adhesive bonding of the materials is satisfactory insofar as:

– the glued joint works by shearing in its plane;

– there are no tensile strain stresses (for example linked to secondary moments that are inherent to purity defects) in the joint.

4

Techniques for Repairing Civil Engineering Works

4.1. Repair of concrete structures

General information

The repair techniques outlined in the following apply to structural or non-structural damage.

Some are already relatively old and well proven (for example the glued metal plates technique dates from the 1960s), while others are still in full development (*bonded pultruded* plates) and require specific assessment procedures (technical advice, ATEX, etc.).

Recently, two standards have entered into play to regulate these various modes of repair.

4.1.1. *The glued metal plates technique*

4.1.1.1. *Principle of the technique*

The reinforcement of concrete structures using the technique of glued sheets consists of overcoming the insufficiencies of resistance that result from degradation of the structure or from an underdimensioning of the latter through the adherence of metal plates on the concrete surface.

This process, which dates from the 1960s, and which is the fruit of the labor of Mr. L'Hermite, Mr. Bresson and Mr. Theillout, is now very well mastered.

Figure 4.1. *Example of floor reinforcement*

4.1.1.2. *Regulations*

This process is the topic of the following texts:

– Issue No. 6 of STRESS: "Technique for repairing and reinforcing concrete structures";

– ITBTP Annals of 1990: "Reinforcement and repair of structures: design and operation";

– ITBTP Annals of 1992: "Repair and reinforcement of buildings and structures";

– NFP 95-105 standard (planned).

4.1.1.3. *Principle for sizing of reinforcements*

4.1.1.3.1. Operation of plated concrete

The preferred mode of operation for plated concrete is shear transmission of the bonding plane of stresses from the concrete structure to the metal reinforcements.

Any other mode (flexion or compression) should therefore be avoided.

As a result, the steel–glue–concrete combination will have a calculated resistance of the weakest material; so in this case, concrete (insofar as it has the lowest surface shear strength).

This provision is valid for the cold calculation of reinforcements (when hot, the bonding epoxy resin is usually the weak point of the combination).

4.1.1.3.2. Magnitude of material characteristics

The value of the bonding plane shear stress was determined experimentally and corresponds to the distribution for a 3 mm thick sheet, as shown in Figure 4.2.

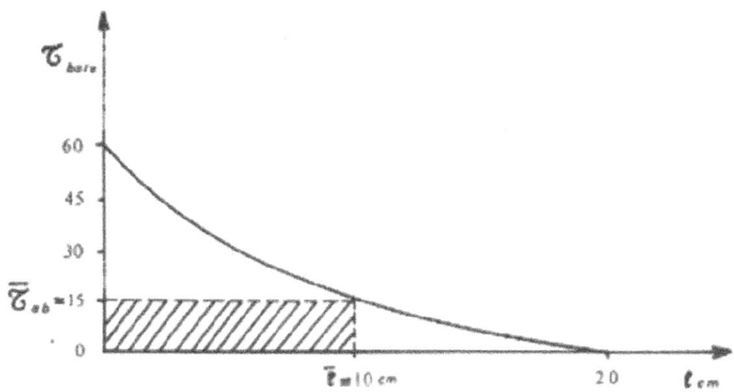

Figure 4.2. *Document by J. Bresson*

This shear stress value should be tested prior to the bonding operation.

4.1.1.3.3. Method for calculating the reinforcement

Method for calculating longitudinal reinforcements under bending forco

This involves, for example, the reinforcement of a beam or a slab that is subjected to bending by the loads applied to it. The reinforcing plates are glued to the underside of the beam.

The justifications shall be carried out in accordance with the BAEL and Eurocode 2 requirements, taking the following modifications into account:

– for, service limit state (SLS) a minority coefficient is applied to sections of internal steels in the structure (Ki) and to sections of the plates (Ke). The values used are:

- $Ke = 1.2 - 0.08 \times e_a$
- $Ki = 0.46 + 0.08 \times e_a$;

– for ultimate limit state (ULS), no minority coefficients are applied;

– the limiting constraint (applied to ELS) of glued plates is given by:

$$\sigma_e \leq 0.47 \times f_e$$

(this coefficient of 0.47 makes it possible to take the local bending of the sheet into account as defined below).

When stacking sheet metal, the forces taken up by the sheet closest to the concrete are:

– 0.66 × F for a stack of two plates;

– 0.5 × F for a stack of three plates.

F = total force taken up by the stack.

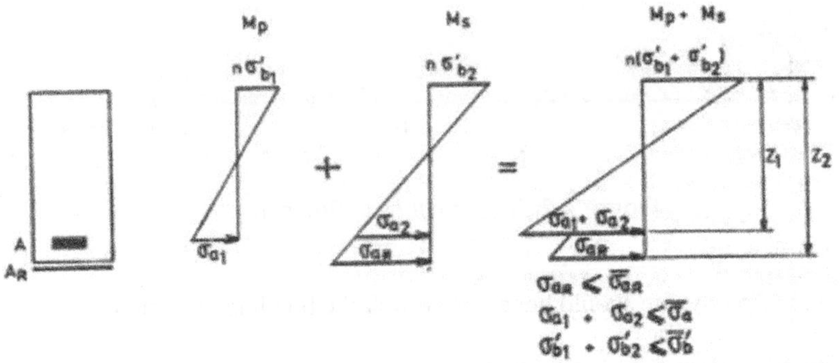

where:

– Mp is the bending moment generated by permanent loads;

– Ms is the bending moment generated by operating costs;

– A is the section of the reinforcement in place in the concrete;

– A_r is the section of the reinforcement.

The resulting equations show the limit of reinforcement possibilities.

The difficulty of having different strength reinforcements on the same structure is another criterion to consider: reinforcing steel of 240 MPa and steel set in place of 240, 400 or 500 MPa.

The shear stress in the glue joint is defined by:

$$\tau = T \cdot S / b \cdot I$$

We must then ensured that the calculated stress remains less than the permissible stress by bonding.

This principle was complemented by M. Theillout in his study. He was interested in the operation of glued sheets that straddle a crack in order to limit the opening of the latter.

Tests carried out at the LCPC made it possible to answer the following questions:

– what is the local behavior of the sheets in the vicinity of the crack?

– what is the distribution of forces in the plates if they are stacked one on top of the other?

– what is the distribution of deformations between the structure's internal steels and the glued plates?

– what are the stresses that cause the concrete plate to peel off?

The results of this study are as follows:

– in the vicinity of the crack, there is a local flexion in the plate.

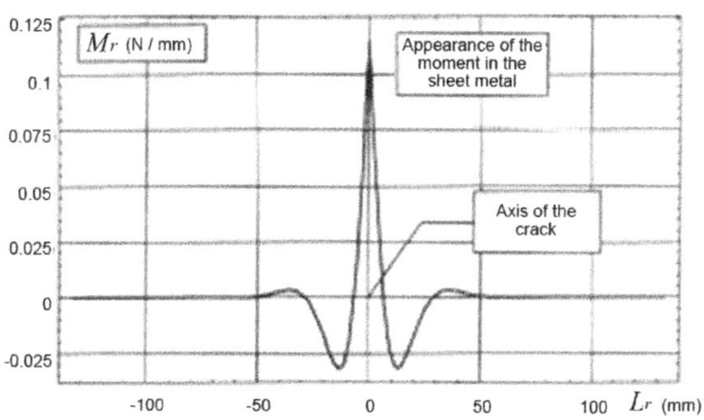

Figure 4.3. *Bending moment in sheet metal (according to J. N. Theillout)*

This local flexion is at its maximum in the line of the crack and decreases very rapidly as one moves away from it.

It generates tensile stresses, which are at their maximum on the outer fiber of the sheet.

The evolution of this maximum stress is summarized in Figure 4.4 as a function of the length of overlap.

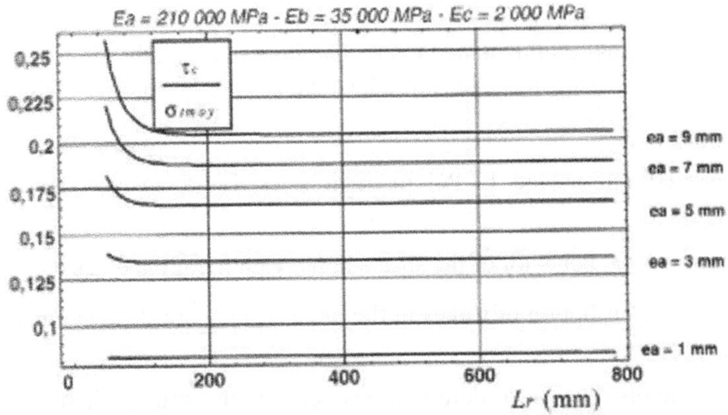

Figure 4.4. *Evolution of the maximum shear stress as a function of overlap length (document by J. N. Theillout)*

This results in plastification by bending the plate.

This arrangement led J. N. Theillout to introduce a bending index, which is the ratio between the average stress applied and the stress exerted on the most stressed fiber;

– In the case of sheet metal overlay, the distribution of forces taken up by each sheet is not identical;

- for 2 plates of the same section, the one closest to the concrete takes up two-thirds of the forces,

- for 3 layers of the same section, the one closest to the concrete takes up half the forces, with the remainder divided equally between the two others;

– the experiment carried out on beams stressed in flexion tends to show that the plates deform more than the internal steels. A strain–stress diagram must therefore be considered for a concrete structure that is subjected to bending of the type proposed by J. N. Theillout:

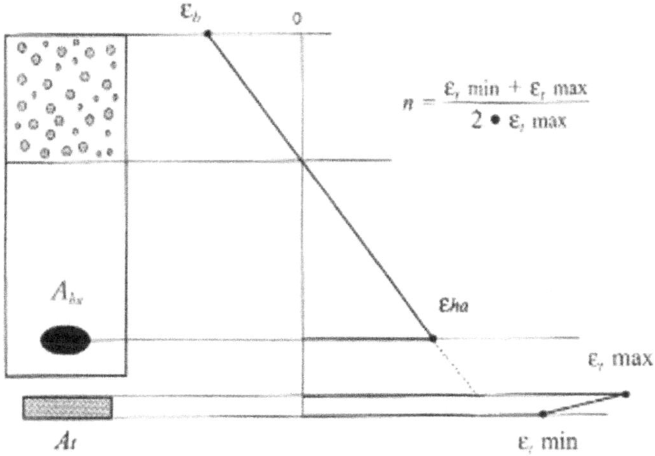

Figure 4.5. *Deformation diagram*

Method for calculating transverse reinforcements under shear stress

The reinforcement of a beam against shear force is done by gluing plates onto the flanks of the beam.

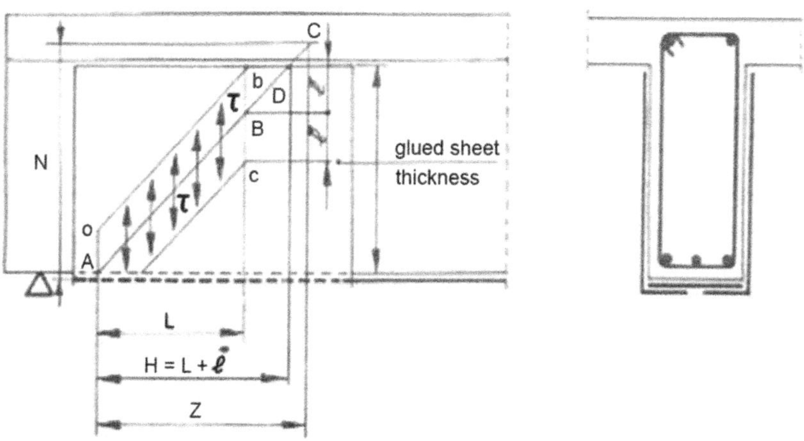

Figure 4.6. *Operating diagram according to J. Bresson*

4.1.1.4. *Implementation of reinforcements*

4.1.1.4.1. Preparation of the support

As the principle of reinforcement is based on bonding, it is essential to ensure that the latter is carried out under optimal conditions.

The concrete surface must therefore be prepared in such a way as to eliminate all the non-bonding parts and make it perfectly flat to ensure the bonding of a relatively rigid element.

The various techniques that can be used are listed in ITBTP Annals No. 62 from June 1976 ("Preparation of concrete surfaces and steel substrates for structural bonding").

Prior to preparation of the support, all the parameters that may have an influence on the bonding must be carefully analyzed, in particular:

– soil;

– oxides;

– hydrocarbon binders;

– carbon black (after a fire, for example);

– oils;

– fats;

– moisture (dryness test);

– soft roe and rough areas;

– presence of curing agent, demolding;

– bubbling.

Two are of particular significance:

– chiselling or bushing with manual or pneumatic tools;

– dry or wet sandblasting.

The use of patching mortar should be avoided since the mechanical properties of the bonding of mortar to concrete are limited and are often lower than those required in the glue joint. This should not exceed 20% of the bonding surface and the ends of the plates should be avoided.

4.1.1.4.2. Bonding of plates

The gluing of plates generally involves the following operations:

– implementation of a bonding primer to improve adhesion between the concrete and the adhesive. This primer penetrates through the porosity of the concrete and allows the powdery bases to be fixed;

– application of the bonding agent. This is mainly epoxy resin under technical advice or technical inquiry;

– since these resins are particularly sensitive to atmospheric conditions (particularly humidity and high temperatures), special attention should be paid to the environment during the bonding phase. Adhesion tests for the adhesive on concrete may be considered at this time;

– implementation of reinforcing steel consisting of S235 steel plates (steels of higher grade should be avoided: plastification of the sheet after detachment may occur). These sheets are sandblasted and prepared in the factory in order to improve their adhesion;

– pressurization of the plates (clamps, cylinders, etc.) to obtain a minimum stress level of around 4 kPa;

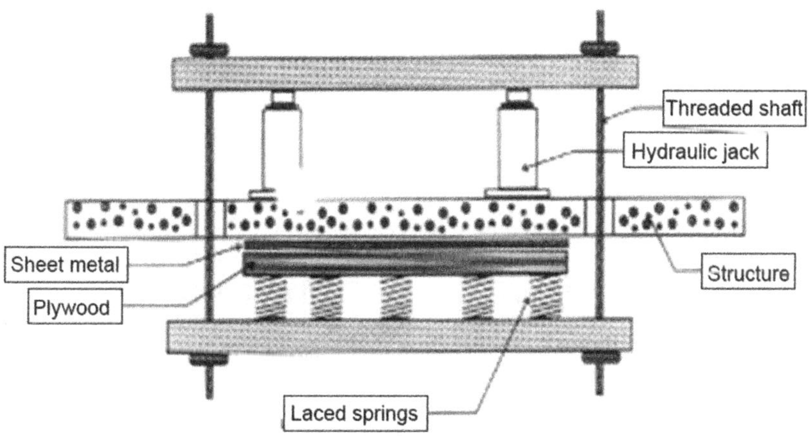

Figure 4.7. *Example of pressurization of plates*

– testing of reinforcement by adhesion tests and loading tests;

– possible protection against corrosion or fire.

Figure 4.8. *Bonding of metal plates (document by SIKA)*

4.1.1.5. *Field of application*

This reinforcement technique has the following advantages:

– relatively low cost;

– improved operation through advances in glue formula.

However, it has considerable disadvantages:

– delicate implementation (jacking, etc.);

– poor knowledge of reinforcement behavior in aggressive atmosphere (waste water plant) or saturated with moisture (tanks);

– lack of results on earthquake resistance;

– additional protection required for heat resistance.

4.1.2. *The technique of glued composite fabrics or plates*

4.1.2.1. *Principle of the technique*

This technique is identical to the technique for glued metal plates but substituting the metal with composite materials, which can be found in two forms:

– fabrics (mainly carbon fibers, fiberglass, aramid fibers in an epoxy matrix);

– pultruded plates (also based on carbon fibers).

Research is currently under way in Europe. Examples include:

– the study on the Polystal ring by Bayer in Germany;

– study on Parafil in Great Britain;

– development of Arapree in the Netherlands;

– use of several processes in France (TFC by Freyssinet, etc.).

Material	Carbon			Glass		Kevlar 49/ epoxy	Boron/ epoxy
	HR T 300/ epoxy	HR T 800/ epoxy	HM M 50/ epoxy	E/ epoxy	R/ epoxy		
Resistance							
Longitudinal traction (MPa)	1,600	2,940	1,080	1,030	1,380	1,380	1,300
Longitudinal compression (MPa)	1,500	1,570	830	550	660	280	2,500
Transverse tension (MPa)	50	60	45	41	41	41	61
Transverse compression (MPa)	120	270	–	138	138	138	202
Interlaminar shear (MPa)	65	100	60	55	55	55	67
Module							
Young's longitudinal E_x (GPa)	132	162	225	45	5.2	72	200
Young's transverse E_y (GPa)	10	10	7	12	13.8	5.5	18.5
Shear G (GPa)	5	5	4.5	4.4	4.5	2.1	5.59
Poisson coefficient ν_{xy}	0.35	0.34	0.3	0.25	0.25	0.34	0.23
Percentage of fibers by volume	60	60	60	60	60	60	60
Density relative to water	1.57	1.6	1.66	2	2	1.38	2

Table 4.1. *Mechanical characteristics of composite fabrics*

4.1.2.2. *Regulations*

In the absence of a standardizing text, these provisions are subjected to the following procedures:

– technical advice (ATEC) for materials for which the specifications have been validated by the CSTB;

– technical appraisal of experimentation (ATEX) for the others.

4.1.2.3. *Principle for sizing of reinforcements*

This is based on the BAEL and Eurocode 2 requirements, taking into account the law of behavior of composite materials used for reinforcement.

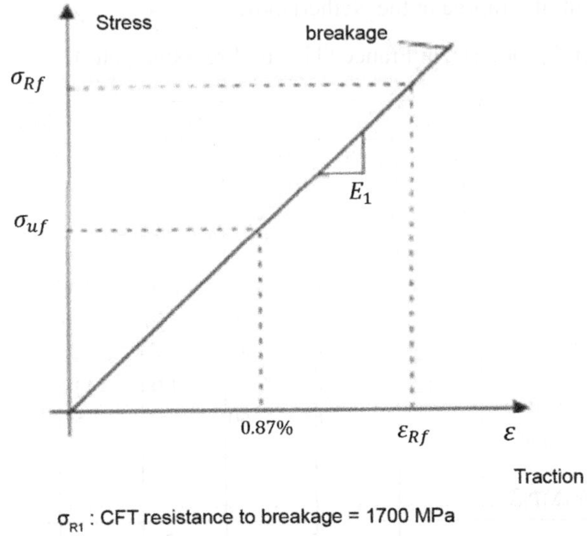

Figure 4.9. *Example of a behavior law for CFT (document by Freyssinet)*

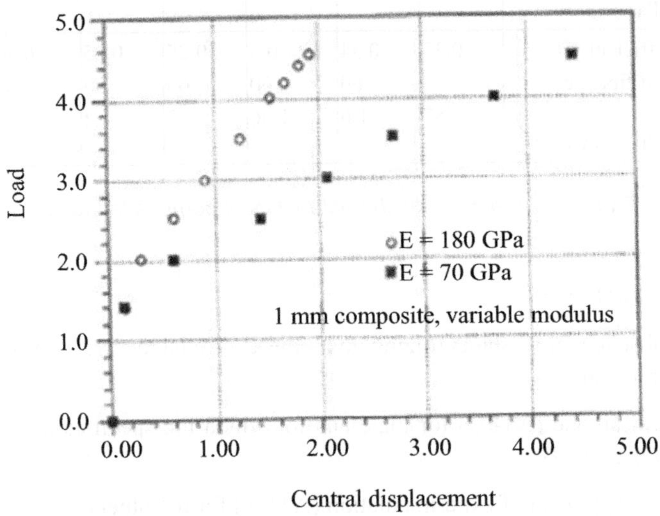

Figure 4.10. *Example of the behavior law for a UD complex (carbon-epoxy)*

Concrete composite bonding is dependent on shear stress, which is usually:

– 1.5 MPa for SLS;

– 2.0 MPa for ULS.

4.1.2.3.1. Method for calculating longitudinal reinforcements under bending force

As mentioned above, the diagram of the state of stresses is shown in Figure 4.11 (this principle can be applied to reinforced concrete structures but also to prestressed concrete structures):

– Flexion in the service limit state:

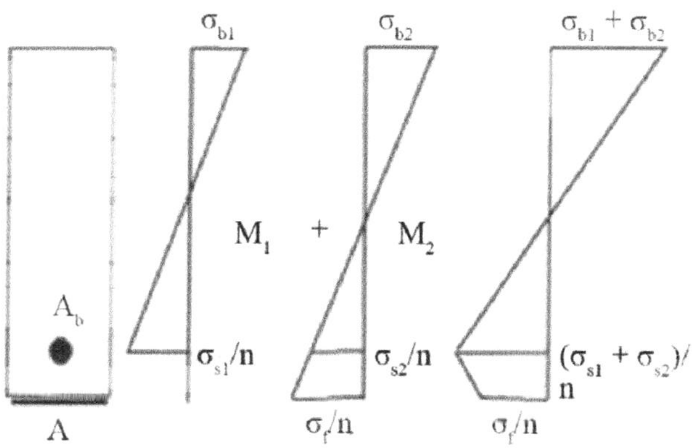

Figure 4.11. *Stresses diagram*

Constraint testing is performed when:

- $\sigma_{s1} + \sigma_{s2} \leq \sigma_s$,

- $\sigma_{b1} + \sigma_{b2} \leq \sigma_b$;

- $\sigma_f \leq \sigma_{sf}$;

where:

- σ_b is the compression stress of concrete: $0.6 \times f_{cj}$;

- σ_s is the permissible tensile stress in the steels in place: $\sigma_s \leq$ fe relative to the selected degree of cracking;

- σ_{sf} is the permissible stress in the reinforcing material (for example 550 MPa for CFT).

– Flexion in the ULS:

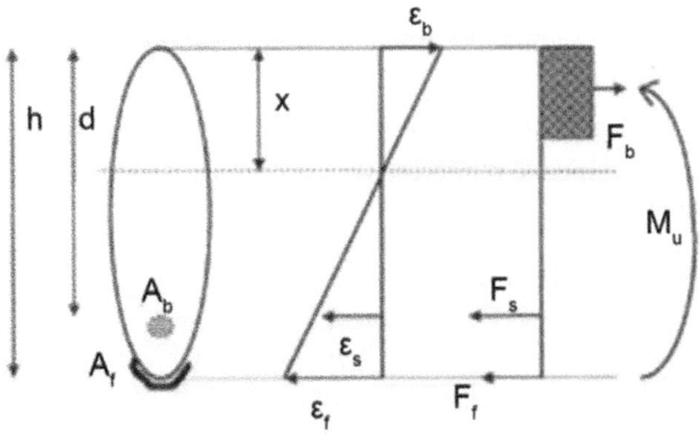

F_f, force to be taken up by the CFT, is deduced from the equilibrium of the section

$M_u = F_s.(d\text{-}x/2) + F_f.(h\text{-}x/2)$

$F_h = F_s + F_f$

Figure 4.12. *Deformation and forces diagrams*

– Entrainment stress:

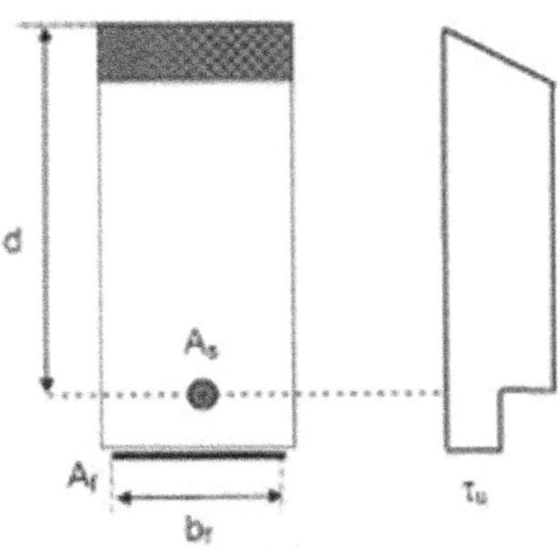

$$\tau_u = \frac{V_u}{b_f d} \cdot \frac{F_f}{F_s + F_f}$$

Figure 4.13. *Entrainment stress diagram*

For prestressed concrete structures, reinforcement by composite materials involves applying a verification class that is directly inferior to that envisaged during initial design.

As a result, the reinforcement checks at the SLS are only carried out in class 2 or 3 for prestressing of structures that were initially designed to be in class 1 or 2.

For retaining structures, it should be ensured that this provision complies with requirements in the CCTG Booklet 74.

– Experimental results:

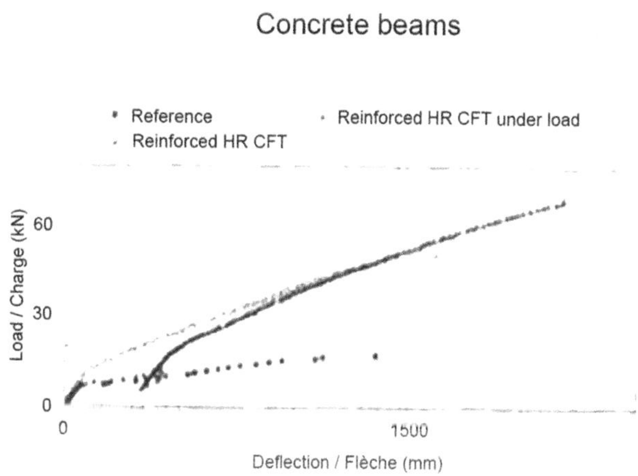

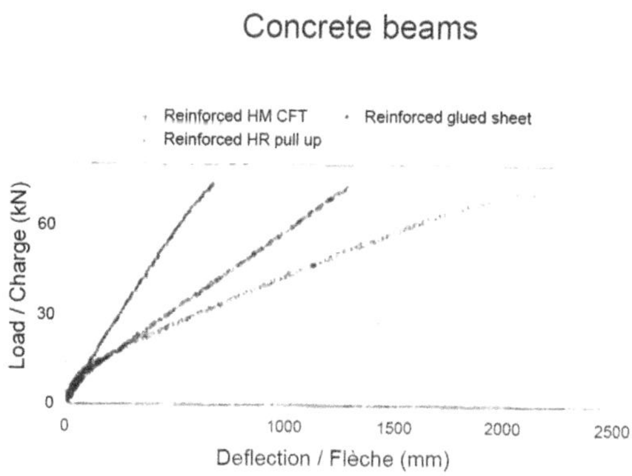

Figure 4.14. *LCPC document with CFT*

A comparison table was also drawn up between the reinforcements by bonded metal plates and CFT (Table 4.2).

Reinforcement	Units	Steel (E 24-3)	Bidirectional CFT
Density	kg/m^3	7.8	1.8
Thickness	mm	3	0.28
Weight	g/m^3	23,400	500
Longitudinal Tensile Strength	MPa	240	2,540
Longitudinal Tensile Modulus	GPa	200	160
Longitudinal Failure Load	kN/cm	7.2	7.1
Transverse Failure Load	kN/cm	7.2	3

*All properties reported for dry fibers.

Table 4.2. *Comparison between reinforcement by bonded metal plates and CFT*

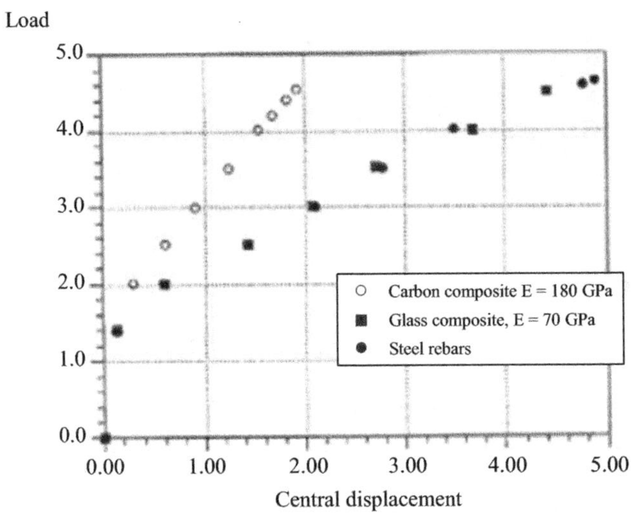

Figure 4.15. *Comparison of reinforced concrete with 1 mm RPF and traditional methods*

4.1.2.3.2. Method for calculating the containment of concrete posts

This involves improving the compressive strength of a concrete post by wrapping closed circular strips of composite fabric around it.

Thus, the gain obtained from the BAEL and Eurocode 2 parabola-rectangle diagram is shown in Figure 4.16.

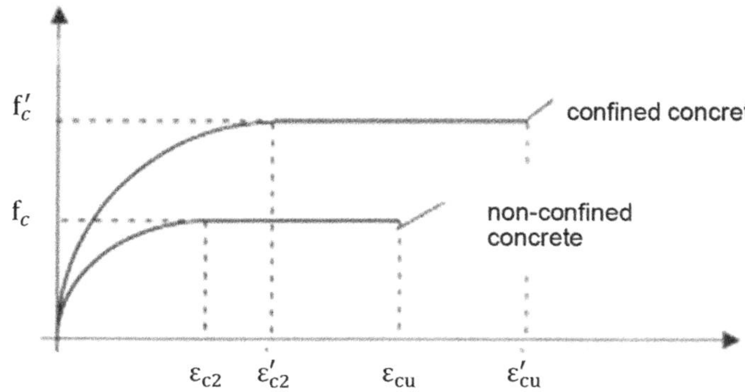

$$f'_c = f_c + 4\,p$$

$$\varepsilon'_{cu} = \varepsilon_{cu}[1 + 20\,p/f_c]$$

$$\varepsilon'_{c2} = \varepsilon_{c2}[f'_c/f_c]^2$$

where p is the confinement pressure of the CFT

Figure 4.16. *Confined parabola-rectangle diagram*

4.1.2.4. *Implementation of reinforcements*

The various steps involved in the implementation are substantially identical to those mentioned for glued metal plates, namely:

– preparation of the support;

– impregnation of the surface with bonding resin;

– application of dry fabric or pultruded plate;

– support to ensure bonding.

4.1.2.5. *Field of use*

This repair process is advantageous in the following cases:

– resistance to corrosion (except in the cases mentioned above);

– resistance to chemically aggressive agents;

– mechanical behavior;

– ease of implementation.

The main disadvantages are as follows:

– heat resistance;

– atmospheric stresses during bonding.

4.1.3. *The technique of additional prestressing*

4.1.3.1. *Principle of the technique*

The two processes mentioned above use structural reinforcements, which can be described as passive as per the definition put forward by the Eurocode 2 and NFP 95-105 "Repair and reinforcement by additional passive reinforcements".

The additional prestress involves applying a force to the structure to be reinforced – the cables needed for applying this force are only vectors.

This prestressing is called additional because it involves increasing the loading capacity of an existing structure.

The main difficulty of the method arises from the fact that the structure to be repaired is cracked and was not originally designed to enable the setup and tensioning of prestressing frames. Consequently, the prestressing should be done on the exterior of the structure to be reinforced.

Figure 4.17. *Example of additional prestressing*

4.1.3.2. *Regulations*

The main regulatory references are the following:

– Issue no. 5 of STRESS;

– ITBTP Annals no. 501: "Repair and reinforcement of buildings and structures";

– standard NFP 95-104.

4.1.3.3. *Principle for sizing of reinforcements*

The principle behind the calculation essentially comes from:

BPEL Appendix 7 ("Concrete external prestress"):

– the possibility of replacing cables;

– easy access to anchors;

– prohibited use of bare external cables (except under the conditions in section 2.3 of BPEL appendix 7.

– justification at the ULS of the stability of the shape if the only link between the prestress and the structure is at the level of the anchors;

– STRESS specifications to keep in mind:

- the structure must remain accessible for monitoring and maintenance;

- the preliminary diagnosis must validate the technical feasibility of the process.

Implementation of additional prestressing must be designed in such a way as to integrate:

– the treatment of cracks;

– modification of the static diagram of the existing structure;

– maintenance of additional cabling.

To calculate the additional cabling needed, one must do so according to the BPEL or Eurocode 2 requirements, if possible, and try to respect the initial static diagram of the structure.

For a circular structure such as a reservoir, the prestressing should be carried out on the outside of the basin after the reinforced concrete bosses have been set up. The sizing and number of bosses and the plotting of spindles should make it possible to avoid causing parasitic forces within the structure, such as:

– ovalization;

– vacuum thrust at the openings.

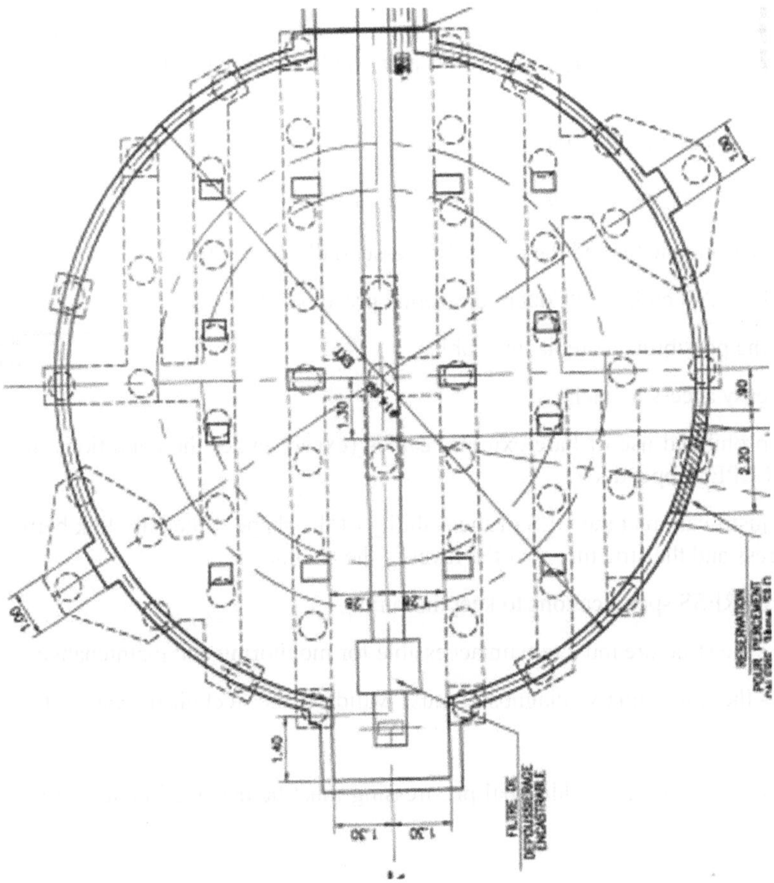

Figure 4.18. *Example of reinforcement of a sugar silo*

4.1.3.4. *Implementation of reinforcements*

4.1.3.4.1. *Materials*

– Prestressing reinforcements: These are identical to those used for new structures and must be included in the list of steels approved by the CCTG Issue 4, Title II. These are generally strands for longitudinal reinforcements and bars for clamps and stirrups.

The protection against corrosion of reinforcements is ensured by:

– greasing of coated steels;

– galvanization;

– use of stainless steels.

In all cases, the protection is implemented at the factory.

– Anchors: these are approved for external prestressing, in particular due to their positioning on the outside, so the parts are protected and can withstand ambient aggression.

– Ducts: the reinforcements are placed in ducts over their entire length. These are:

- corrosion-resistant metallic tubes (galvanized or epoxy-coated) with a thickness above 2 mm;

- flexible tubes made of plastics (HDPE type). The Circular 99-53 recommends using NF tubes from groups 2 (drinking water) and 4 (industry). The duct fittings are then included in the standard product range.

4.1.3.4.2. Example of implementation

Reinforcement of structures through external prestressing requires a precise implementation methodology. It must include the following elements:

– detailed implementation of the prestress (geometric limit values corresponding to calculations);

– installation of structures to be cast in place or to be fixed to the structure (for prefabricated elements, for example) such as bosses, diverters, end elements that receive anchoring of cables, etc.;

– drilling (trying to avoid damaging existing reinforcements);

– installation of ducts and anchors;

– definition of the cable tensioning program;

– tensioning of bars;

– injection of cracks;

– injection of ducts;

– protection of elements.

Figure 4.19. *Example of beam reinforcement*

Figure 4.20. *Silo reinforcement*

4.1.3.4.3. Inspections

The quality assurance plan must define the means for controlling external prestress. In particular, it must incorporate:

– suitability tests: the capacity of the material and materials required to properly achieve the intended structure;

– blank test;

– checking the efficiency of the injection (measuring gauge on cracked sections);

– verification of the effectiveness of the prestressing;

– verification under load.

4.1.4. *The shotcrete technique*

4.1.4.1. *Principle of the technique*

The technique consists of projecting concrete on the wall to be strengthened by discharge through a nozzle. This is carried out by adherent thin layers because of the projection pressure.

This technique can be applied to the projection of:

– mortar (aggregate less than 4 mm);

– sand concrete (absence of gravel and smaller dosage of cement than in mortar).

Shotcrete is an interesting option to consider when it is not possible or if it is very difficult to form the structure to be repaired. Its normal use is generally as follows:

– filling of voids (molded wall to be replenished after failure of concreting, etc.);

– rejointing of masonry;

– production of a protective coating (protection against fire by coating steels, etc.);

– structural reinforcement (reinforcement coating and increase in the inertia of the structure, etc.);

– tunnel vault, retaining wall (such as Berlin walls), etc.

4.1.4.2. *Regulations*

The main regulatory references are the following:

– Issue no. 6 of STRESS;

– NFP 95-102 standard;

– NF EN 14487-1 and NF EN 14487-2 standards on the execution of shotcrete;

– NF EN 14488-1 to 7 standard for shotcrete tests;

– NF EN 934-5 standard for shotcrete admixtures.

4.1.4.3. *Principle for sizing of reinforcements*

To calculate the steel sections required for reinforcement, reference must be made to BAEL and Eurocode 2. Tests carried out at the LCPC showed that the reinforced structure could be considered to have a monolithic behavior.

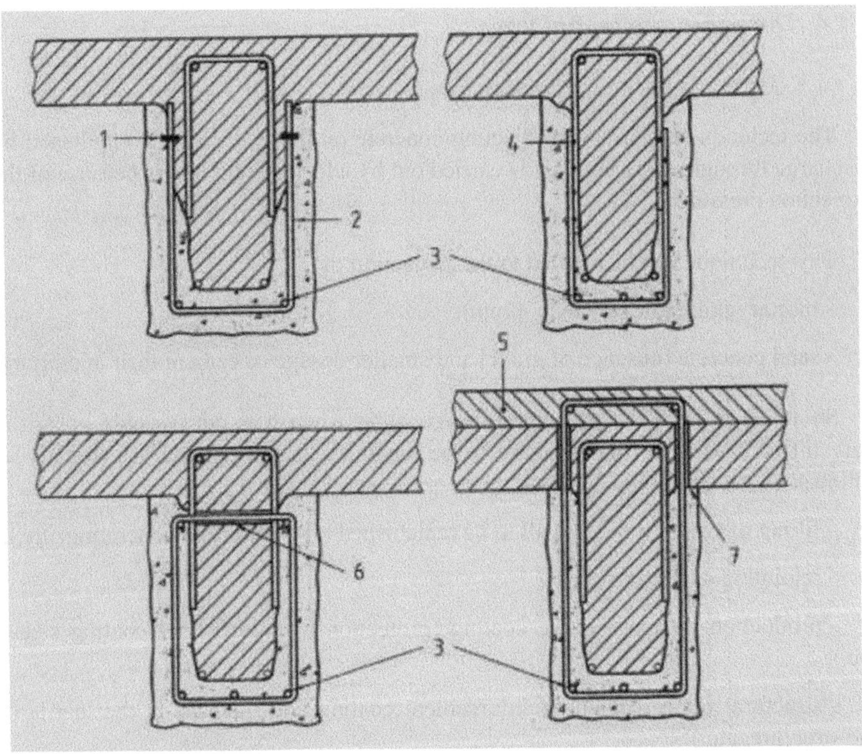

Figure 4.21. *Reinforcement 1: reinforcement for anchoring new reinforcement in existing concrete (connectors); Reinforcement 2: shear force transverse reinforcement (usually welded mesh); Reinforcement 3: longitudinal flexion bending reinforcements (usually bars); Point 4: welding points between existing and new reinforcement (solution to be justified); Point 5: concrete poured at the top of the beam (increase in inertia); Points 6 and 7: various drilling for the passage of new reinforcements*

This should result from a prior technical study, which must validate the adequacy of the solution to the problem posed at each stage of the project.

For structures to be reinforced, existing reinforcements are commonly ignored in the final calculation of the structure.

4.1.4.4. *Implementation of reinforcements*

4.1.4.4.1. Projection procedures

There are two methods for using shotcrete:

– dry spraying:

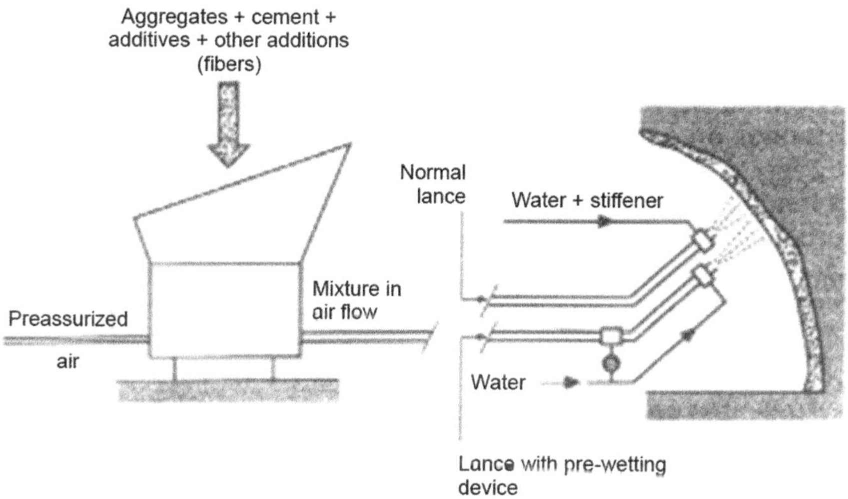

Figure 4.22. *Principle of dry spraying*

Its use is advantageous in the following cases:

– small-scale project;

– site with difficult access;

– site where the distance between the projection machine and the lance exceeds 100 m;

– site where concrete strength is high.

– wet spraying:

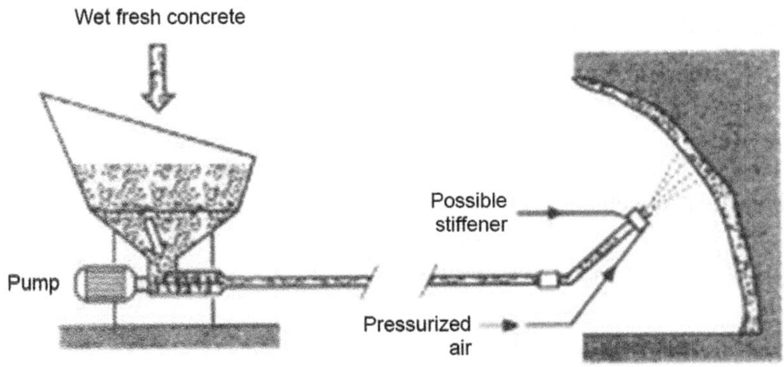

Figure 4.23. *Wet method with dense flow*

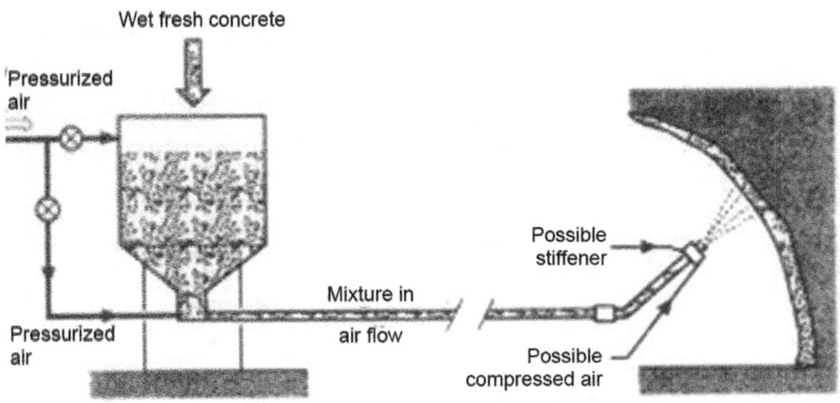

Figure 4.24. *Wet method with diluted flow*

This is mainly used in the following cases:

– high-yield sites;

– work sites where safety is a limit on the emission of dust (tunnel, galleries, pipeline, etc.).

The advantages and disadvantages of the two processes are summarized below:

	Advantages	Disadvantages
Dry method	High compactness Strong mechanical resistance Excellent adhesion to the substrate	Quality of the mix depending on the projector Very significant rebound losses Production of dust
Wet method	Better quality of the mixture Limited rebound losses	Low horizontal transport distance (about 100 m for diluted flow and 150 m for dense flow) Lower mechanical resistance Poorer adhesion Coating of delicate reinforcements

Table 4.3. *Advantages and disadvantages of the dry and wet methods*

4.1.4.4.2. *Materials*

Shotcrete is made up of:

– aggregates conforming to standards NFP 18-542 and XPP 18-540. For the dry method and in order not to disturb the water dosage, the water content must be less than 6 % by weight;

– hydraulic binders conforming to standard NF EN 197-1 for common cements. For structures in aggressive environments, refer to P 18-011 and EN 206-1;

– water that is compliant with XPP 18-303;

– additives and additions (accelerators, superplasticizers, etc. or slag additions, fly ash, etc.).

The reinforcements are common reinforced concrete frames that are certified by the AFCAB. Fibers can optionally be used after being tested for suitability.

The composition of shotcrete must meet the following criteria:

– resistance criterion: the cement dosage should be chosen according to the target objective;

– sustainability criterion according to the aggressiveness of the environment;

– compactness criterion;

– finesse criterion: fine element content greater than 17% of the mixture by weight.

The AFTES recommendations, which are based on processes, are complementary to the above. In particular, they target:

– a ratio of sand to sand plus gravel such that:

- by wet method: $0.70 \leq S/(S+G) \leq 0.90$;

- by dry method $0.60 \leq S/(S+G) \leq 0.80$;

– a cement dosage such as that mentioned in the table below:

Target use of the shotcrete	Cement content of concrete in place* (kg/m³)	Cement dosage of the mixture (kg/m³)	
		Dry method	Wet method
Masonry repair mortar	500	400	500
Surface repair	350	280	350
Structural repair Structural strengthening	450	360	450
*The value indicated is an average value of the active cement (clinker equivalent) for the entire thickness of the projected layer (>2 cm).			

Table 4.4. *Cement dosage at the manufacturing of the shotcretes according to their target use and the cement content of the concrete in place Adapted from the document by AFTES*

4.1.4.4.3. Preparation of the support

This depends on the nature of the repair:

– for structural repairs, all degraded materials should be removed. The surfaces are then blown and moistened slightly before projection. *Resin-based adhesive products should be avoided* as they could impair the adhesion of concrete (see LRPC Aix en Provence test);

– for repairs to masonry joints, degraded elements should be pitted until the material is healthy;

– for repairs to masonry cladding, the preparation is done by sandblasting the structure and then moistening before projection.

4.1.4.4.4. The projection of concrete

The lance is positioned at about 0.50–1.50 m from the support depending on the projection speed. As an indication, the velocity at the lance outlet for the dry method is around 100 m/s.

The thickness of a layer is generally less than 10 cm.

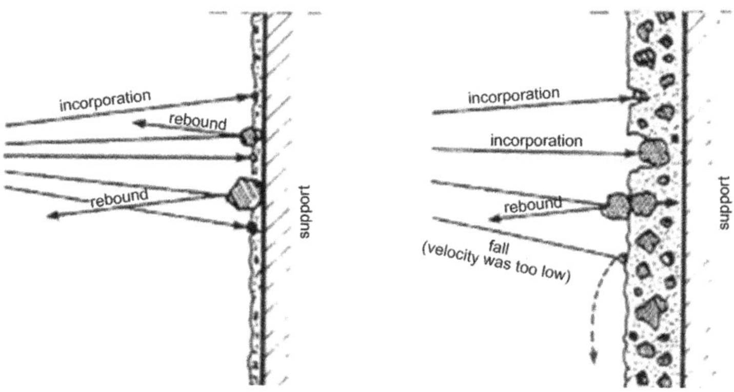

Figure 4.25. *Principle of concrete incorporation*

For the dry method, the optimum value of a layer is around 5 cm.

For the wet method, it is possible to go up to 7 cm.

4.1.4.4.5. The tests

The standard includes the following:

– Studies that consist of:

- granulometric analysis of aggregates;
- formula of the mixture to be sprayed;

- tests on specimens (tensile strength, compression);
- compatibility of the elements.

– Suitability tests for:

- supplies;
- effectiveness of treatment;
- composition of the mixture;
- consistency measurements (cone collapse, penetrometer, etc.);
- core specimens;
- bonding;
- thicknesses implemented.

– Inspection tests primarily consisting of:

- adhesion tests (performed in a laboratory);
- core samples for traction compression tests.

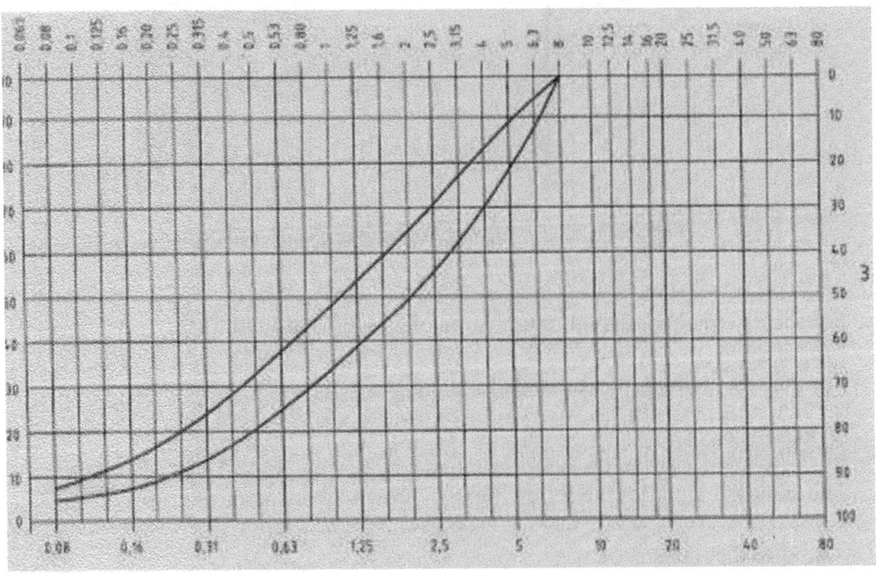

Figure 4.26. *Example of a particle size distribution according to P 95-102*

Figure 4.27. *Shotcrete*

4.1.5. *Repair of superficially degraded concretes*

4.1.5.1. *General Information*

4.1.5.1.1. Principle of the technique

This technique applies to all reinforced, unreinforced and prestressed concrete structures that have superficial defects without showing structural pathologies.

Any degradation is regarded as superficial if it leads to a reduction in the quality of protection of concrete at the surface and up to a depth of a few centimeters but it does not jeopardize the stability of the structure.

Three types of repairs can be carried out:

– products that are based on hydraulic binders and on modified hydraulic binders;

– mixed products;

– products based on synthetic resins.

The principle then covers the implementation of products after preparation of the support:

– surface treatment (patching);

– injection and treatment of cracks.

4.1.5.1.2. Regulations

The applicable texts are as follows:

– *EN 1504-1*: Products and systems for the protection and repair of concrete structures. Definitions, requirements, quality control and evaluation of conformity. Part 2: Surface protection systems for concrete;

– *EN 1504-3*: Products and systems for the protection and repair of concrete structures. Definitions, requirements, quality control and evaluation of conformity. Part 3: Structural and non-structural repair;

– *EN 1504-4*: Products and systems for the protection and repair of concrete structures. Definitions, requirements, quality control and evaluation of conformity. Part 4: Structural bonding;

– *EN 1504-5*: Products and systems for the protection and repair of concrete structures. Definitions, requirements, quality control and evaluation of conformity. Part 5: Concrete injection;

– *prEN 1504-6*: Products and systems for the protection and repair of concrete structures. Definitions, requirements, quality control and evaluation of conformity. Part 6: Anchoring of reinforcing steel bar;

– *prEN 1504-7*: Products and systems for the protection and repair of concrete structures. Definitions, requirements, quality control and evaluation of conformity. Part 7: Reinforcement corrosion protection;

– *EN 1504-8*: Products and systems for the protection and repair of concrete structures. Definitions, requirements, quality control and evaluation of conformity. Part 8: Quality control and assessment and verification of the constancy of performance;

– *ENV 1504-9*: Products and systems for the protection and repair of concrete structures. Definitions, requirements, quality control and evaluation of conformity. Part 9: General principles for the use of products and systems;

– *EN 1504-10*: Products and systems for the protection and repair of concrete structures. Definitions, requirements, quality control and evaluation of conformity. Part 10: Site application of products and systems and quality control of the works.

4.1.5.1.3. Principle for sizing of reinforcements

As this is not a structural reinforcement, this issue is not discussed in this chapter.

4.1.5.1.4. Implementation of reinforcements

Materials

As seen previously, the three groups are as follows:

– Products based on hydraulic binders: these are mortars or concretes composed of binders and aggregates. This category is subdivided into two further categories according to whether or not the cement is modified by addition of synthetic polymers (acrylic or vinyl resin type);

– Products based on cement and active organic polymers (mixed products): this family is dominated by epoxy-cement systems. The three predosed components are mixed at the time of use only, with both resin components on one side (base + hardener) and cement and fillers on the other side;

– Products based on synthetic resins: these consist of sand, polymers and mineral fillers. The most common products are epoxy resins, polyurethanes, polyesters (see section on synthetic materials).

The choice of products to be used depends on the results from the preliminary study. The most common *arrangements for patching* are as follows:

– hydraulic mortars after application of an epoxy bonding layer;

– epoxy mortars after attaching an epoxy bonding layer;

– polyurethane mortars on polyurethane bonding layer.

Characteristics	Hydraulic binder	Epoxy resins	Polyurethane resins
Dry support adhesion	+	+++	++
Wet support adhesion	+++	+	–
Cracking of the support	+	+	+++
Passive effect	+++	–	–
Thermal compatibility	+++	+	+
Penetration of liquids	++	+++	+++

Table 4.5. *Table of characteristics for different mortars*

Preparation of the support

Preparation for patching

– removal of degraded and non-adhering concrete;

– cleaning of the surface and removal of concrete that is contaminated by chlorides or carbonation;

– check the pH of concrete around the reinforcing bars of steels over 1–2 cm;

– protection and possible replacement of reinforcements;

– implementation of the patching.

Figure 4.28. *Failure of repair due to incorrect preparation of support*

To prepare the surface of the concrete before patching, the following techniques can be used:

– hydro-demolition (usually for high thicknesses and to preserve steels);

– hydroblasting;

– very high pressure water scouring;

– pneumatic hammer stitching.

It is better to avoid:

– bush hammering and chiselling;

– combustion engine concrete breaker;

– chemical scouring;

– mechanical planning;

– sanding.

To clean the support, the following methods are usually implemented:

– pressure washing;

– brushing, suction, blowing.

To protect steels, the following are usually done:

– scouring (Sa2 degree);

– installation of an anticorrosion coating if the repair mortar cannot be used immediately or if the coating is insufficient;

– use of a slurry based on hydraulic binders just before patching.

Preparation for injection and treatment of cracks

This is identical to the procedure described above. The mechanical characteristics of the concrete that is in contact with the injected material should be those of the structure in its healthy parts.

4.1.5.1.5. Implementation of grouting

The NF EN 1504-3 standard defines four classes of products based on their performances:

– structural repairs are class R3-R4;

– non-structural repairs are class R1-R2.

Implementation of the patching

The implementation steps are detailed in the standard. However, we highlight the importance of the following aspects:

– the attachment layer (roughcasting, resin, etc.);

– good mixing of the components;

– compliance with the thicknesses to be applied according to the product specifications;

– protection of hydraulic mortars (curing, etc.).

Implementation of injection and treatment of cracks

The different treatment processes depend on the nature of the crack and the activity.

A distinction is usually made between:

– caulking based on putty, bridging through a coated canvas and surface impregnation (mainly for faichings);

– the injection.

The capacity of the product to inject in a crack depends on its opening (cement grout injections are reserved for those with the bigger openings and resin fluid for smaller openings).

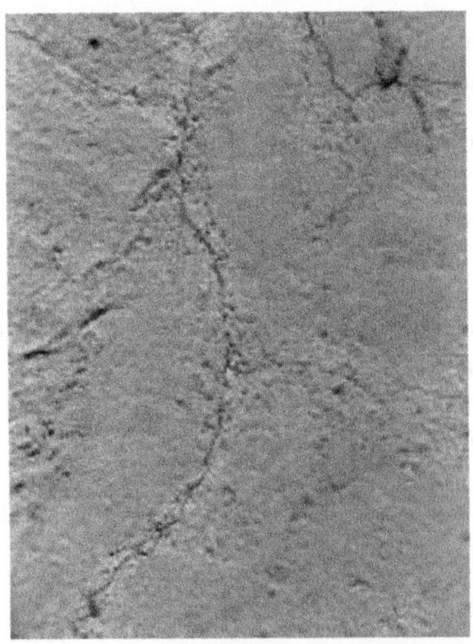

Figure 4.29. *Shrinkage cracks before injection*

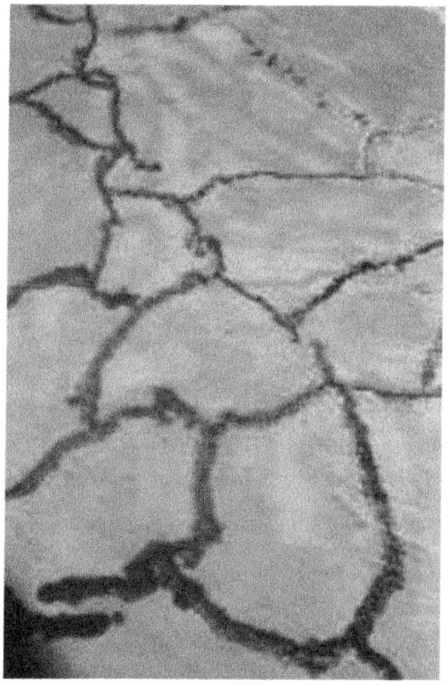

Figure 4.30. *Injected cracks*

4.1.5.1.6. Sealing of passive reinforcements in concrete

This method is used:

– to replace existing reinforcements that have been subjected to damage;

– to add new reinforcements to reinforce the structure.

Regulations

– FD P 18-823: "Special products for hydraulic concrete constructions – Sealants based on hydraulic binders or synthetic resins – Recommendations for the dimensioning of reinforcing bar seals in concrete";

– Standard NF EN 1881: "Products and systems for the protection and repair of concrete structures – Test method – Testing of anchorage seals using the tearing method".

This technique requires prior knowledge of:

– the characteristics and condition of the reinforcements already in place;

– the characteristics of the concrete (mechanical and chemical);

– the actions to be taken into account;

– the aggressiveness of the environment.

The implementation of the new reinforcement in the structure can be done in the following ways:

– by covering the bars, which requires removing the steel already in place over a certain distance;

– by sleeving or coupling bars through a process that involves Technical Notice;

– by welding under the conditions of the EN 1992-1-1.

Table 3.4: Permitted welding processes and examples of application

Loading case	Welding method	Bars in tension[1]	Bars in compression[1]
Predominantly static (see 6.8.1 (2))	flash-welding	butt joint	
	manual metal arc welding and metal arc welding with filling electrode	butt joint with $\phi \geq 20$ mm, splice, lap, cruciform joints[3], joint with other steel members	
	(AC2) metal arc active welding (AC2)	splice, lap, cruciform[3] joints & joint with other steel members	
		-	butt joint with $\phi \geq 20$ mm
	friction welding	butt joint, joint with other steels	
	resistance spot welding	lap joint[4] cruciform joint[2, 4]	
Not predominantly static (see 6.8.1 (2))	flash-welding	butt joint	
	(AC2) manual metal arc welding (AC2)	-	butt joint with $\phi \geq 14$mm
	metal arc active welding	-	butt joint with $\phi \geq 14$mm
	resistance spot welding	lap joint[4] cruciform joint[2, 4]	

Notes:

1. Only bars with approximately the same nominal diameter may be welded together.
2. Permitted ratio of mixed diameter bars $\geq$ 0,57
3. For bearing joints $\phi \leq$ 16 mm
4. For bearing joints $\phi \leq$ 28 mm

Table 4.6. *Table 3.4 from EN 1992-1-1*

4.2. Protection of concrete structure

4.2.1. *Cathodic protection of reinforcements*

4.2.1.1. *Principle of the technique*

The aim of cathodic protection is to stop the corrosion process before mechanical risks reach a high level. It is usually applied to maritime structures to protect all types of metallic materials.

This method of protecting concrete reinforcements against corrosion involves lowering the electrical potential all along the reinforcement to a point called the protection potential.

To do this, an electrical current passes from the coating to the reinforcement as shown in Figure 4.34.

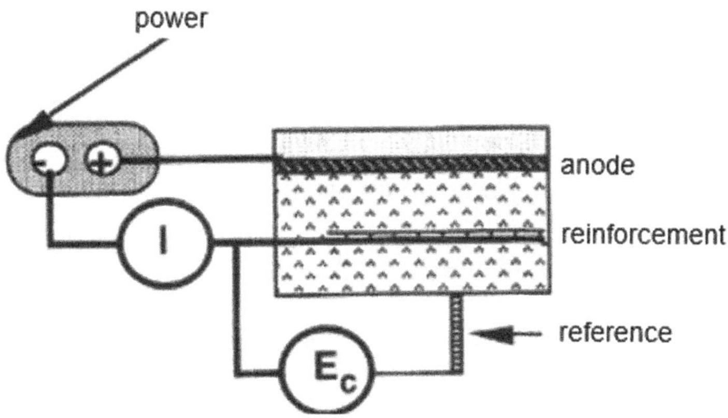

Figure 4.31. *Principle of cathodic protection*

There are several different types of processes. The most common are:

– a process with primary anodes placed in grooves cut by sawing into concrete with a spacing of about 7.50–30 m. Secondary anodes are arranged in longitudinal grooves every 20 cm (maximum) and are protected by a vinylester resin. Its main defect is that it is not durable;

– a process with anodes to be coated with concrete: titanium trellis anode covered with concrete (usually shotcrete for vertical surfaces);

– a process with an anode in conductive coatings.

4.2.1.2. *Regulations*

This method of reinforcement protection for reinforced and prestressed concrete is the main topic of the following regulation:

– Standard EN 12-696: Protection of steels in concrete.

4.2.1.3. *Principle for sizing of reinforcements*

This method is not a reinforcement technique but a protection one.

4.2.1.4. *Implementation of the protection*

In order for cathodic protection of reinforcements to be possible, certain parameters should be taken into consideration:

– the reinforcements in place must have a sufficient degree of conservation (section of steels);

– only local repairs are considered;

– the zones where the reinforcements are corroded must be delineated;

– the chloride content of the concrete must be accurately determined to avoid further deterioration of the structure;

– the thickness of the steel coatings that directly affect the flow of electric current must also be specified;

– there should be no screen between the reinforcement and the anode.

The anode setup must ensure cathodic protection. In particular, its life span must be compatible with that of the project (with potential maintenance work).

– if the anode is embedded in the concrete, the current density must be in accordance with that of the design and must not exceed the limit values;

– if the anode is embedded in a conductive coating (organic or hot-sprayed metal), the anode is used as an anode with an imposed current.

4.2.1.4.1. Application example for a tank

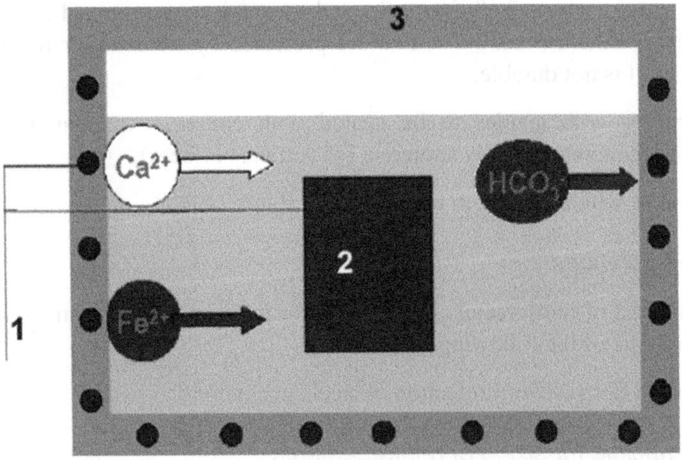

Figure 4.32. *Diagram of corrosion of reinforcements in a water reservoir*

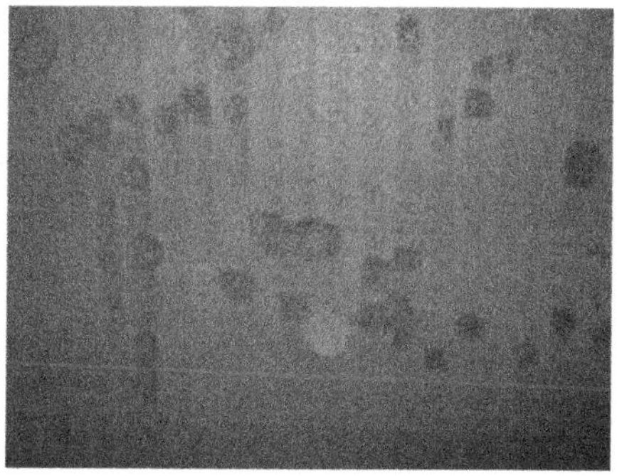

Figure 4.33. *Trace of steel corrosion on a tank cover dome*

Implementation of cathodic protection

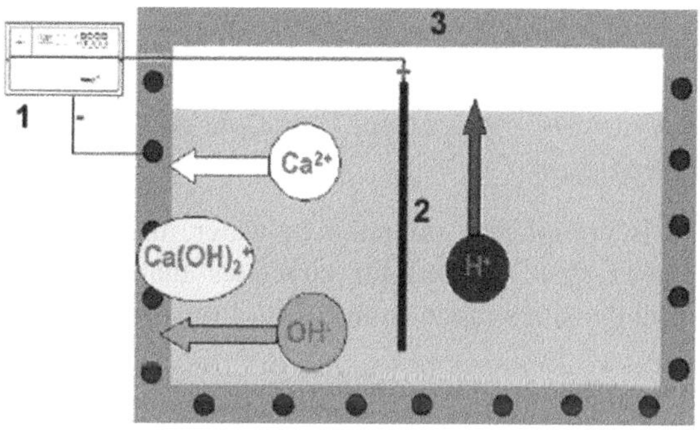

Figure 4.34.

After protection has been added

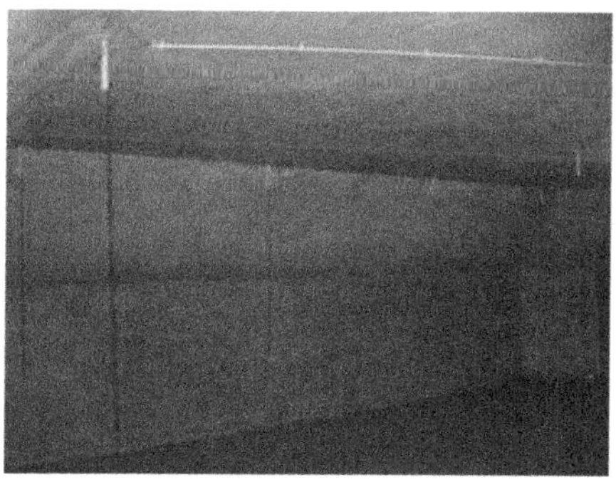

Figure 4.35. *Tank cover dome after implementation of a cathodic protection*

It is then necessary to:

– assess corrosion protection: two criteria are generally considered:

- the measurement of a negative potential (around 850 mV relative to a reference electrode Cu/$CuSO_4$);

- a potential difference of 100 mV after removal of the rectifier.

– regular checkups.

4.2.1.5. *Density of current required for "cathodic prevention" and "cathodic protection"*

Typical current densities:

– for cathodic prevention, this ranges from 0.2 to 2 mA/m^2;

– for cathodic protection, this varies from 2 to 20 mA/m^2.

4.2.1.6. *Density of current required for "cathodic protection" for underground and submerged structures*

If the concrete is completely saturated in water, the current density may be considerably less than that required for concretes that are exposed to external atmospheres. Typical densities range from 0.2 to 2 mA/m^2 for new structures (prior to initial corrosion).

For structures that are unsaturated or already corroded, the current densities are identical to those of structures exposed to air, with values of up to 20 mA/m^2.

A stack effect can also be created between the fully immersed part and the overhead part of the structure. A higher current density will then have to be applied to the immersed part.

4.2.1.7. *Special case for prestressing steels*

These steels are particularly sensitive to hydrogen (embrittlement). EN 12-696 therefore recommends not to subject them to potentials that are more negative than –900 mV with respect to Ag/AgCl/KCl 0.5 M to avoid this embrittlement by hydrogen.

For cathodic protection by currents imposed in the protection of prestressed elements that are already corroded, the safety potential limit must be determined by laboratory tests.

4.3. Underground recovery

4.3.1. *Principle of the technique*

The need for underground recovery may be linked to a number of factors:

– foundation design failure (lack of geotechnical study or erroneous conclusions, etc.);

– failure to take account of parasitic effects (for example negative friction or dissymmetrical thrust for piles, etc.);

– implementation failure (misrecognition of seat level, compressible terrain, plug failure for piles, poor concreting, etc.);

– unplanned ground movements;

– construction of a structure that is deeper than its neighbor;

– etc.

Depending on the nature of the foundations (superficial or deep), underground recovery techniques may differ.

In all cases, it involves transferring the foundation loads to a support that satisfies the construction requirements in terms of stresses and strains.

4.3.2. *Regulations*

The following texts are applicable to foundation structures:

– Issue 62 Title V of the CCTG for civil engineering works;

– DTU 13.1 and 13.2 for building structures;

– NFP 95-106: Engineering structures – repair and strengthening of concrete and masonry constructions – specifications pertaining to the structure foundations;

– approved specifications for specific techniques.

4.3.3. *Principle for sizing of reinforcements*

The principle results from the application of the texts cited above. In general, it is assumed that existing degraded foundations are no longer able to fulfill their role.

4.3.4. *Implementation of reinforcements*

4.3.4.1. *Recovery of surface foundations*

4.3.4.1.1. Recovery by alternating passes

The foundations must be lowered to the level of the new seat floor by inserting concrete blocks under the foundations. This is done through primary and secondary studs so as not to destabilize the superstructures.

This technique, which requires human presence at the excavation, remains limited to small-scale recoveries.

Figure 4.36. *Example of underground recovery on shallow foundations*

4.3.4.1.2. Recovery by piles

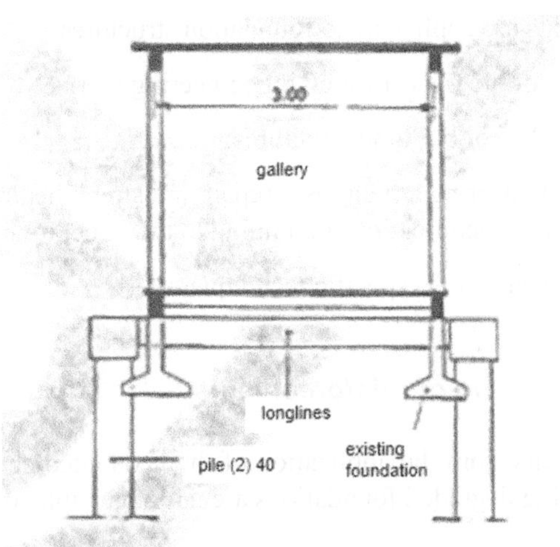

Figure 4.37. *Example of recovery by piles*

This principle involves taking over the superstructures in place through a new system of deep foundations. The load descents are brought onto the piles through a network of longlines.

The most used deep foundations in these cases and within the limit of their bearing capacity are micropiles.

4.3.4.1.3. Recovery by specific techniques

Among these recovery techniques, there are different methods to reinforce the ground under the foundations:

– to increase the bearing capacity of the surface foundations;

– to reduce settlement.

The main techniques are as follows:

– *Ground injection technique*:

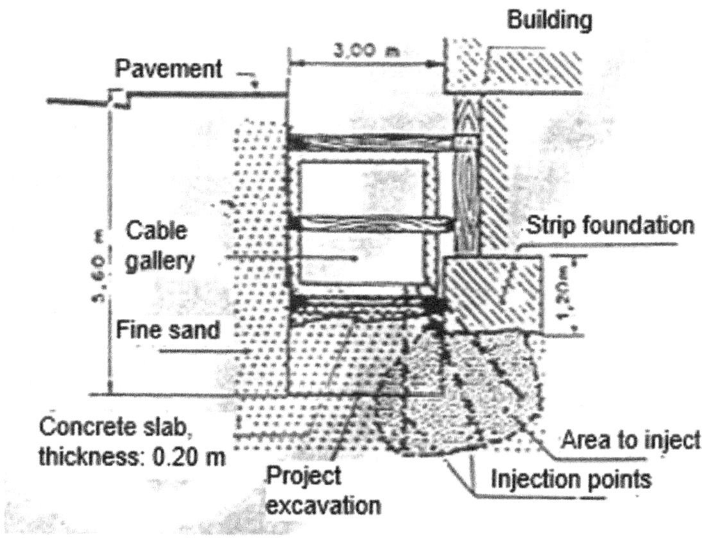

Figure 4.38. *Example of underground recovery by ground injection*

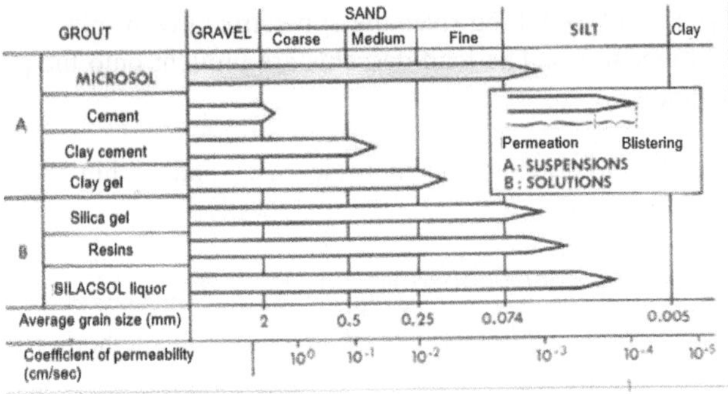

Figure 4.39. *Example of application of injection grout*

– *Jet grouting technology: (NF EN 12-716)*

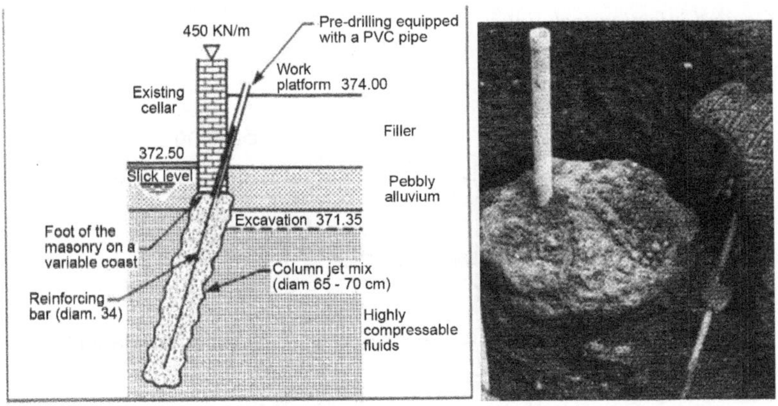

Figure 4.40. *View of a jet grouting column*

The implementation through jets of elements or ground-cement structures may be applied to temporary or permanent structures with different purposes. These include:

– the construction of foundations for new structures;

– the underground recovery of existing foundations;

– the implementation of low permeability screens;

– the construction of supporting structures or load-bearing structures;

– the completion of works that are ancillary to other geotechnical works;

– the reinforcement of a ground block.

4.3.4.2. *Recovery of deep underground foundations*

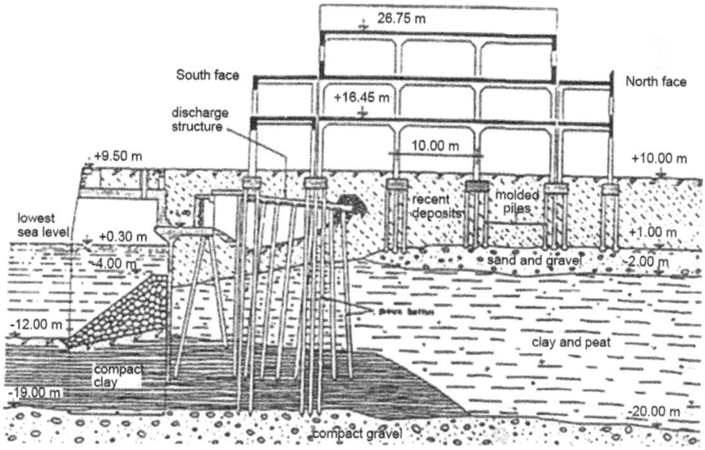

Figure 4.41. *Foundations of the Le Havre maritime station*

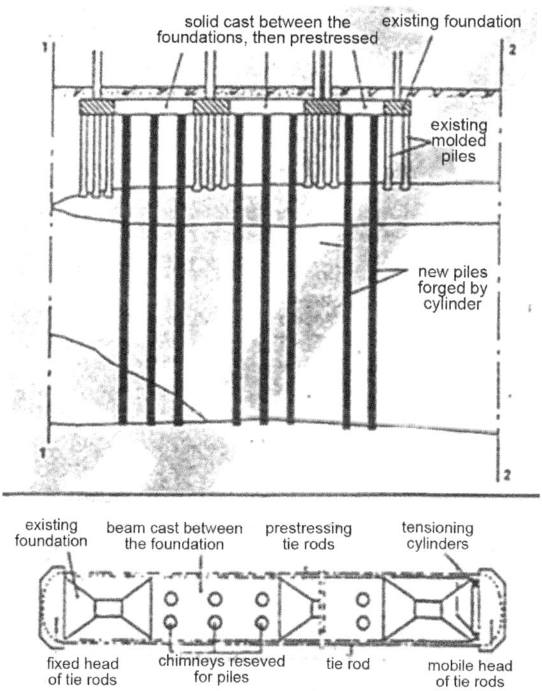

Figure 4.42. *Implementation of new piles*

The pathology of the construction is related to the anchoring of piles on a bank of sand and gravel that is not very thick.

M. Freyssinet's underground recovery involved forging new concrete piles up to the compact gravel layer to create a network of prestressed *longlines and to hydraulically jack the superstructure.*

5

Inspection and Maintenance of Structures in the United States: Methodologies

The principles we just mentioned apply more or less naturally to all countries that use the European standards, as we explained in the Introduction.

Some countries outside the Eurocode area of application have also chosen to integrate these into their national standards.

This is the case, for example, in some Francophone African countries (Ivory Coast, Senegal, Cameroon, etc.) and some Anglophone African countries (South Africa, Nigeria, etc.).

The national appendices allow certain local specificities to be taken into account (cyclonic zone in Madagascar, for example).

Thus, the previous chapters we described above are usually valid for most countries.

This chapter focuses on the different inspection methodologies practiced in the United States for civil engineering structures. In most cases, they are very similar to those used in Europe.

5.1. Engineering structures

5.1.1. *General information*

As mentioned above, inspection and maintenance operations must already be considered during the design phase of a civil engineering structure. The duration of use of a bridge can only be assured if a maintenance and inspection program is put in place as soon as construction is decided.

5.1.2. *Regulations*

Generally, the design and construction of bridges must meet the American Association of State Highway and Transportation Officials (AASHTO) requirements.

Inspection and maintenance rules are often published by State, for example:

– the "Manual for Bridge Assessment" from the AASHTO (Washington, DC);

– the New York State Inspection Guide: "Fundamentals of Bridge Maintenance and Inspection";

– the "National Bridge Inspection Standards" (NBIS);

– etc.

There are different ways to inspect a structure depending on the State, but the minimum requirements are those from the Federal Highway Administration, which recommends, for example, a frequency of 2 years between inspections. The latter must be carried out by a team comprising a certified project manager.

In some cases, the biennial inspection is supplemented by an in-depth inspection every 4–5 years.

Referring to the New York State Inspection Guide, it has been shown that the frequency of inspection visits of every 24 months for all motorway bridges over 20 feet has made it possible to reduce the degradation of structures.

Thus, New York State defined four intervals:

– Type 1: biennial inspection. This is recommended for all motorway bridges as it is a routine inspection. It should also apply to new or renovated structures within 60 days before opening to the public.

– Type 2: annual inspection. This applies to the most degraded structures and falls between two biennial inspections.

– Type 4: no periodicity. These are bridges that are closed to traffic during renovation or reconstruction. It may also include inspection of temporary structures.

– Type 5: special inspection. These are specific inspections that are not part of routine inspections and are not entered into the project file database. It is often used for complex or exceptional structures at the request of the head of the department (Deputy Chief Engineer).

– Type 3: These are inspections prior to renovations. It is more akin to a diagnosis of structures.

5.1.3. *Human resources*

The inspection team usually consists of:

– a project leader (team leader or TL) who must have engineering experience (validated by the State of New York, for example) or 5 years' experience in inspecting civil engineering structures;

– an assistant (assistant team leader or ATL).

The role of the TL is to ensure that the bridge has been fully inspected and that the submitted inspection report is in accordance with the requirements of the State specifications (for example, the New York State Bridge Inspection Manual for the State of New York) or the Federal Highway Administration.

The role of the ATL is to inspect and take all measurements of structures and different pathologies within the framework of a quality assurance plan validated by a Quality Control Engineer.

The inspection program should be drawn up by an engineer who has at least the following qualifications:

– be a certified and registered engineer or have at least 10 years' experience in bridge inspections;

– successfully passed a federal training graduation exam in inspection of structures.

5.1.4. *Material resources*

The equipment required for carrying out inspections includes:

– a hard hat, safety jacket, traffic control signals;

– a mason's hammer to probe concrete;

– a positive or negative scale or nacelle;

– a camera;

– a measuring device, a caliper, an ultrasonic measuring device;

– fatigue crack measurement equipment;

– orange spray paint for marking deteriorated elements;

– cones for road traffic;

– etc.

5.1.5. *The inspection report*

The structural elements are classified based on their level of degradation according to the scale presented in Table 5.1.

<table>
<tr><th>Index</th><th>Evaluation of the condition</th><th>Level of renovation</th></tr>
<tr><td>1</td><td>Structures with severe mechanical damage with immediate risk of ruin.</td><td rowspan="3">Structural renovation program to be planned.</td></tr>
<tr><td>2</td><td>Structures with severe mechanical disturbances between indices 1 and 3.</td></tr>
<tr><td>3</td><td>Structures with serious degradation or malfunctions.</td></tr>
<tr><td>4</td><td>Structure between indices 3 and 5.</td><td rowspan="6">Routine maintenance.</td></tr>
<tr><td>5</td><td>Structures with low damage and operating correctly.</td></tr>
<tr><td>6</td><td>Structure between indices 5 and 7.</td></tr>
<tr><td>7</td><td>Structures in good state of conservation and operation.</td></tr>
<tr><td>8</td><td>Structures in very good state of conservation and operation.</td></tr>
<tr><td>9</td><td>Structures in excellent state of conservation and operation.</td></tr>
<tr><td>N</td><td>Unknown or not inspected.</td><td></td></tr>
</table>

Table 5.1. *Degradation index of structures*

Routine maintenance usually involves:

– cleaning the deck, supports, de-icing salt areas;

– maintaining the rainwater drainage system;

– cleaning the expansion joints;

– checking the concrete seals.

Corrective or restorative maintenance consists of:

– resealing the expansion joints;

– repainting metal structures;

– recoating;

– increasing rainwater networks.

In a renovation operation, the inspection report is a reference document.

In this case, the inspection team is asked to rate not only the state of conservation and operation of the bridge but also to measure the level of degradation of each structural element.

This provision will make it possible to set priorities within the context of the renovation operation.

For example, there is a specification stating that bridge beams that have lost more than 20% of their geometric characteristics require their bearing capacity to be recalculated.

The "New York State Inspection Guide" recommends a signalization of priorities in the form of flags:

– red flag: lack of stability or imminent risk of loss of stability of a critical element in the main structure;

– yellow flag: condition of damage that could lead to a significant risk before the next inspection. This flag can also be used to indicate an imminent loss of stability of a secondary structural element;

– safety flag: immediate risk of danger to vehicles or pedestrians without risk of loss of stability or collapse of the structure. These can also be used for bridges closed to traffic.

If the red flag and the safety flag are up, the inspector must specify "Prompt Interim Action" in the report if immediate action is to be taken.

5.1.6. *Points to look out for*

Type of damage	Probable causes	Severity index	Repair solution
Crack in the coating (for over 25% of the surface area)	Delamination Flaking	3	Replacement Local repairs
Crack in the coating (small surface area)	Delamination Flaking	5	Cleaning Treatment of cracks in the pavement and the deck Resurfacing
Waterproofing damage in the deck (over 75% of the surface area)	Absence of surface seal Deterioration of the sealing layer	3	Repair or installation of a waterproof coating

Localized defective waterproofing of the deck	Absence of surface seal Deterioration of the sealing layer	5	Bridge cleaning Treatment of cracks in the pavement and the deck Resurfacing
Damage to roadway seal	Mechanical wear Runoff of rainwater	2	Repair or replacement of the joint
Deterioration or deformation of supports	Ice Wear, fatigue	3	Repair or replacement
Piles. Generalized cracking of concrete, efflorescence, flaking	See Chapter 3, section 3.1	2	Repair, reinforcement Timely repair (see Chapter 4, section 4.1.5) Shotcrete (see Chapter 4, section 4.1.4)
Piles. Cracking of concrete, efflorescence, flaking where the surface area does not exceed 25% of the total area	See Chapter 3, section 3.1	5	Cleaning Timely repair (see Chapter 4, section 4.1.5)
Abutments. Generalized cracking of concrete, efflorescence, flaking	See Chapter 3, section 3.1	2	Repair, reinforcement Timely repair (see Chapter 4, section 4.1.5) Shotcrete (see Chapter 4, section 4.1.4)
Abutments. Cracking of concrete, efflorescence, flaking where the surface area does not exceed 25% of the total area	See Chapter 3, section 3.1	5	Cleaning Timely repair (see Chapter 4, section 4.1.5)

5.2. Storage structures for petroleum products

5.2.1. *General information*

The main principles mentioned above also apply here. The constructive modes are identical and depend very little on the countries in which the structures are built.

The reference standard code is the API 653 code “Tank Inspection, Repair, Alteration and Reconstruction” from the American Petroleum Institute.

Like the DT94 of the inspection UFIP, it mainly covers metal storage tanks but also civil engineering structures such as the foundations.

The main causes of deterioration of the foundations mentioned in API 653 code for storage structures are the following:

– chemical attack from aggressive ground water;

– freeze-thaw cycle;

– sulfate and chloride attacks;

– cracking due to shrinkage and thermal phenomena.

5.2.2. *Inspection procedure*

Appendix C of the API 653 lists the civil engineering works to be inspected (Table 5.2).

TANK IN-SERVICE INSPECTION CHECKLIST			
	Item	Completed ✔	Comments
C.1.1	**FOUNDATION**		
	Measure foundation levelness and bottom elevations (see Appendix B for extent of measurements).		
C.1.1.1	**Concrete Ring**		
	a. Inspect for broken concrete, spalling, and cracks, particularly under backup bars used in welding butt-welded annular rings under the shell.		
	b. Inspect drain openings in ring, back of waterdraw basins and top surface of ring for indications of bottom leakage.		
	c. Inspect for cavities under foundation and vegetation against bottom of tank		
	d. Check that runoff rainwater from the shell drains away from tank.		
	e. Check for settlement around perimeter of tank.		
C.1.1.2	**Asphalt**		
	a. Check for settling of tank into asphalt base which would direct runoff rain water under the tank instead of away from it.		
	b. Look for areas where leaching of oil has left rock filler exposed, which indicates hydrocarbon leakage.		
C.1.1.3	**Oiled Dirt or Sand**		
	Check for settlement into the base which would direct runoff rain water under the tank rather than away from it.		
C.1.1.4	**Rock**		
	Presence of crushed rock under the steel bottom usually results in severe underside corrosion. Make a note to do additional bottom plate examination (ultrasonic, hammer testing, or turning of coupons) when the tank is out of service.		
C.1.1.5	**Site Drainage**		
	a. Check site for drainage away from the tank and associated piping and manifolds.		
	b. Check operating condition of the dike drains.		
C.1.1.6	**Housekeeping**		
	Inspect the area for buildup of trash, vegetation, and other inflammables buildup.		

Table 5.2. *List of civil engineering works to be inspected according to the API 653*

5.2.3. *Points to look out for*

The provisions from European rules may be applied in the absence of any contrary indication from the API 653 code.

Appendices

Appendices

Appendix 1

Examples of Diagnosis on a Drinking Water Storage Structure Based on the CEMAGREF Method

Tower tank. **Overview.**

A1.1. Description of the structure

A1.1.1. *Location*

Figure A1.1. *Aerial view of site*

A1.1.2. *Identification of the structure*

Name of the structure:	MONDOT tower tank
Water agency:	City of SAINTES
Municipality:	Libourne
Structure class (CCTG Issue 74):	B

A1.1.3. *General description of the structure*

It is a tower tank made of up reinforced concrete. It consists of a circular tank resting on annular concrete columns. The tank is supported by the lower belt of the tank and is covered with a thin shell. The reservoir cover dome is coated with a resin-type liquid sealing system. The tank is coated with a waterproofing system.

Foundations	– Not visible
Barrel	– It consists of 12 reinforced concrete annular columns that are each 1.10 m in diameter. – The thickness of the ring is about 18 cm.
Dome of the reservoir	– The reinforced concrete shell is connected to a horizontal slab at the central core. – Its thickness could not be measured.
Reservoir cover dome	– The reinforced concrete shell is connected to a horizontal slab at the central core. – Its thickness is about 9–10 cm.
Equipment or superstructures	– Waterproofing of the upper cupola by liquid sealing systems. – Telephone broadcasting antennas in the center of the cover and on the acroterions. – Metal railings on the periphery of the tank. – Spiral staircase. – Water supply and drain lines and pumps.

Table A1.1. *General description*

A1.1.4. *Size*

– Total height of the structure:	32 m (height of columns: 17 m height of the tank: 15 m)
– Lower diameter:	18.30 m
– Upper diameter:	18.30 m
– Deflection of the lower dome:	Not measured
– Deflection of the higher dome:	Not measured
– Height of railings:	1 m
– Tank volume:	1,000 m^3

A1.1.5. *Terms of use*

Drinking water retention structures.

A1.2. Conditions for development of the structure

Construction of the structure:

– Company:	Unknown
– Consultancy:	Unknown
– Year of construction:	Unknown

Reported incidents:

– None.

Organized supervision of the structure:

– None.

Work done since last visit:

– Not specified.

A1.3. Information relating to the inspection

– Organization responsible for this operation:	n/a
– Previous inspection:	None
– Date of this operation:	June/July 2008
– Constitution of the inspection team:	One team consisting of one civil engineer and one civil engineering technician
– Other participants:	Core drilling company
– Means used:	Visual examination Ferroscan tests Destruction testing
– Atmospheric conditions:	Variable weather
– Other conditions	None
– Particular difficulties encountered	Examination of the outer walls required work on ropes

A1.4. Inspection of the structure

A1.4.1. *Identification*

The tank is located in a rural area.

A1.4.2. *Structure accessibility*

Elements	Description	Location	Notice	Photo no.
Structure accessibility	Structure located in a rural area	–	–	1
Roadway	Asphalt at the road and gravel inside the zone	–	Satisfactory condition	–

Table A1.2. *Accessibility*

A1.4.3. *Superstructures*

Reminder of the CEMAGREF classification

Any visible damage is classified into a category that corresponds to a severity index. There are classes with increasing severity from A to F.

– Class A: structure in good condition without any damage;

– Class B: structure with defects that already existed at the construction with no significant consequences other than aesthetic;

– Class C: structure with defects that are likely to develop abnormally

– Class D: structure with defects that reveal a degradation (three subclasses: D1, D2, D3);

– Class E: structure with defects that reflect a change in the behavior of the structure and affect the lifetime of the structure;

– Class F: structure with defects that indicate the proximity of a limit state and require a restriction of use or decommissioning.

A1.4.3.1. *Tank support columns*

A1.4.3.1.1. Visible defects on structures

The concrete columns that were examined show no visible defects. They could be categorized in *class B* according to the CEMAGREF methodology.

A1.4.3.1.2. Condition of the structures

The columns were subjected to three core drillings for chemical analysis. The results of these analyses are presented in section A1.7. The following interpretation of the results can be made:

1) *Composition of concretes*: The samples taken reveal three mineral matrices of similar composition. This is a concrete with a siliceous-type granular load (between 89 and 98%).

The cement dosages range from 306 kg/m^3 (sample 1) to 265 kg/m^3 (sample 2) and 231 kg/m^3 (sample 3). At the same time, the water contents allow to calculate the W/C ratio for each element. These values range from 0.76 (sample 1) to 0.81 (sample 2) and 1 (sample 3).

The hydration rate values (respectively 19, 18 and 15%) indicate normal setting of the binder.

The compressive strengths of the samples are:

– Sample 1: 9.7 MPa;

– Sample 2: 18.8 MPa;

– Sample 3: 10.3 MPa.

– *Comments on the analyses*: To examine these elements, three observations are proposed:

- the cement dosages are relatively low compared to the dosage recommended in EN 206.1 (not applicable at the time of construction), but it should be noted that issue 74 from March 1983 specifies a minimum dosage of 300 kg/m^3 for any reinforced concrete that is not in contact with water (Article 32);

- the W/C ratios are also very high (between 0.76 and 1). The recommended value is around 0.5–0.55. It is likely that in order to ensure easier implementation of the concrete in the annular space, water was added to the site;

- the values of compressive strengths are low (another consequence of a low binder assay and high W/C ratio). However, this analysis must be qualified because of the geometric dimensions of the cores supplied.

Indeed, NF EN 206.1 determines strength classes from a test on hardened concrete samples with a slenderness of 2, which was not the case for our samples (sample 1, slenderness 2; sample 2, slenderness 1.9 and sample 3, slenderness 1.4).

Moreover, according to NF EN 12504.1, the ratio between the maximum diameter of the aggregates and diameter of the test piece has a significant impact.

As a result, the provided values are for information purposes only.

2) *Products of concrete alterations:* The sought-after products are mainly chlorides, alkalis and carbonation depth;

The chloride ion dosage shows values of less than 0.02% by mass of concrete, or a cement dosage below 0.18 %, the alkaline contents are around 0.10%, the measurements of carbonation depths are as follows:

– Sample 1: exterior: 13 mm interior: 0 mm

– Sample 2: exterior: 0 mm interior: 0 mm

– Sample 3: exterior: 35 mm interior: 0 mm.

– *Comments on the analyses*: These values call for the following comments:

- chloride dosage: The values obtained are lower than the recommended values (0.20% for reinforced concrete);

- alkaline content: Values are also lower than those recommended by the LCPC (3.5 kg/m^3);

- the carbonation depth of 35 mm for a concrete thickness of 18 cm and the coating on sample three measured at around 5 cm should be taken into account for preventive measures;

- for a 1,000 m^3 tank, the compression stress on the columns is around 1.5 MPa and is therefore less than the permissible stress of 0.6 × f_{cj} (6 MPa for the lowest measured value).

A1.4.3.2. *The tank*

A1.4.3.2.1. Visible defects in structures

The tank structure consists of a bottom dome, a cover dome and vertical walls. The bottom dome is supported by columns through a reinforced concrete belt. The cover dome rests on the peripheral sheets through a high belt. There was not any evidence of pathology in the dome or walls. These structures could be placed in *class C*.

The cover dome shows signs of carbonation of the concrete and corrosion of the reinforcements leading to the falling of concrete plates inside the tank. This part of the structure is to be *classified as E*.

A1.4.3.2.2. Condition of the structures

1) Visual and instrumental inspection:

Inspections were carried out with a Ferroscan PS 200 to check the covers of reinforcements inside and outside the tank, as well as point measurements of the phenophthalein carbonation depth.

The values found are summarized in Table A1.3.

Location	Inspection method	Measured coating (mm)		Carbonatation (mm)
Central barrel	Horizontal scanning Vertical scanning	Min: 28 Min: 20	Max: 70 Max: 40	Between 5 and 10
Bottom dome	Horizontal scanning Vertical scanning	Min: 38 Min: 30	Max: 90 Max: 40	Between 5 and 8
Walls	Horizontal scanning Vertical scanning	Min: 30 Min: 35	Max: 40 Max: 80	Less than 10

Table A1.3. *Measured coatings and carbonation thickness*

(The digital outputs of the measurements can be found in the Appendix).

2) Laboratory test on a drilled core:

A coring was carried out on the roof dome. The results of the chemical analyzes can be found in the Appendix. The interpretation of the results is summarized below:

- *Composition of concrete*: The concrete sample has a granular load that is predominantly silica (97.2%).

The cement dosage is 254 kg/m^3. The water content is 9.2% for a W/C ratio of 0.82. The value of the hydration rate (19%) indicates normal setting of the binder. The compressive strength was estimated to be 7.5 MPa.

– *Comments on the analyses*: These values call for the following comments:

- as before, cement dosages are low. The minimum value of 300 kg/m^3 is not achieved. The result is weak mechanical resistance and poor compactness, which may explain the current state of the dome and its carbonation;

- the W/C ratio is exceptionally high (0.82). The currently accepted values are a maximum of 0.5. It is therefore highly probable that in order to ensure easy implementation of the thin shell, water was added to the site. This high water dosage also contributes to the low compactness of the concrete and to the acceleration of carbonation;

- the mechanical resistance value of the concrete is very low. Despite previous comments on the representativeness of the value, and considering the design of the thin shell dome, we advise reinforcement of the latter.

3) *Concrete alteration products:* as mentioned above, the desired products are chlorides, alkalis and the carbonation depth;

The chloride dosage indicates values of 0.09% for the cement dosage, the alkaline content is around 0.10%, the carbonation penetration value is 0 on the sample taken.

– *Comments on the analyses*: These values call for the following comments:

- chloride dosage: the values are below the recommended values;

- alkaline content: the values are below the limit values;

- carbonation depth: the value obtained through coring is not representative of the pathologies identified. Indeed, the corrosion of steel with concrete plates dropping is characteristic of an advanced carbonation state.

A1.5. Summary

Defects found

Defect	Cause	Fixing solution
Construction defects		
Concrete breaking from the roof dome	Poor compactness of the concrete, insufficient cover due to very low (although regulatory) thickness	Heavy repair
Waterproofing problem	Waterproofing of the tank is old and no longer protects the concrete against carbonation	Carry out internal sealing of the tank
Tank support columns: start of carbonation	Protection by highly degraded paint	Implement technical protective coating

Maintenance failure defects		
Not applicable		
Operational defects		
Accumulation of water in the dip on the cover	Sealing of rainwater inlets	Replace the EP runs (if necessary, place them outside if the site classification permits)
Alteration of railings	Oxidation of the tank's metal access ladder	Replacement

Table A1.4. *Nature of defects*

Degree of corrosion:

– Oxidation of the tank guardrails and access ladder at the bottom dome.

Analysis of operating conditions:

– The presence of cover antennas limits the maintenance conditions of the waterproof lining.

Operation of existing measuring devices:

– Not applicable.

Interpretation of observations:

– There are no major structural defects in this structure that could jeopardize its immediate functioning.

– The carbonation of concrete and the level of corrosion of the roof dome reinforcement require rapid action to be taken, otherwise the structure will degrade quickly and definitively.

– The internal sealing of the tank has to be redone, that of the dome seemed to be in good condition.

– The columns and outer walls of the tank should be protected to prevent carbonation.

Evolution compared to the previous detailed inspection:

– No inspection details provided.

A1.6. Conclusion

State of the structure

Part of the structure	Probable causes	Severity index	Possible repair
Surroundings		B	
Tank support columns	Carbonation of concrete	B	Technical painting after purging and local repairs
Bottom dome	Carbonation of concrete and corrosion of steel	C	Implement sealing of the tank
Tank and tank walls (inside)	Carbonation of concrete and corrosion of steel	C	Implement sealing of the tank
Tank walls (outside)	Carbonation of concrete and corrosion of steel	C	Technical painting after purging and local repairs
Cover dome	Carbonation of concrete and corrosion of steel	E	Rework the subsurface with shotcrete after purging. Additional protection against moisture
Overall structure (highest severity level)		E	

Table A1.5. *State of structure*

Proposed work:

Routine maintenance

– Waterproofing of the roof dome and rainwater inlets.

Specialized maintenance

– None.

Major repairs

– See Table A1.5.

Modernization

– None.

Additional investigations to be undertaken

– None.

A1.7. Supplementary material

A1.7.1. *Analysis of concrete*

CONCRETEST® ANALYSIS REPORT

Date received:	20/05/2008	Site of sampling: NR
File reference:	12220/08/5646	
Sampling date:	NR	Sample reference: Coring no. 1
Sampling done by:	NR	Concrete age: NR

WEIGHT COMPOSITION OF CONCRETE		
Parameters	Results	Units
Cement content	13.8	%
Aggregates content	75.8	%
Silicate aggregates	98.2	%
Calcite aggregates	1.8	%
Water content	10.4	%
	100	%

CONSTITUTIONAL PARAMETERS		
Parameters	Results	Units
Cement content	306	kg/m ^{3}
Water content	232	L
W/C ratio	0.76	–
Hydration rate	19	%
Entrained air content estimate	7.0	%

PHYSICAL PARAMETERS			
Parameters		Results	Units
Open porosity	(8)	12.8	%
Visibile density	(1)	2222	kg/m ^{3}
Soaked density	(1)	2350	kg/m ^{3}
Compressive strength	(9)	9.7	N/mm ^{3}
Resistance class		C 8/10	--

CHEMICAL PARAMETERS			
Parameters		Results	Units
Basic parameters			
Fire loss 105°C / 450°C	(2)	***2.60***	%
Fire loss 450°C / 550°C	(2)	***0.25***	%
Fire loss 550°C / 950°C	(2)	***1.11***	%
Soluble silicate content	(5)	***4.45***	%
Na_2 content	(4)	***4.85***	%
MgO content	(4)	***1.92***	%
Al_2O_3 content	(4)	***12.15***	%
SiO_2 content (insoluble)	(4)	***47.6***	%
P_2O_5 content	(4)	***0.212***	%
SO_3 content	(4)	***1.70***	%
K_2O content	(4)	***2.26***	%
CaO content	(4)	***19.8***	%
TiO_2 content	(4)	***0.299***	%
MnO content	(4)	***0.090***	%
Fe_2O_3 content	(4)	***2.23***	%
ZnO content	(4)	***0.026***	%
SrO content	(4)	***0.031***	%
PbO content	(4)	***0.008***	%
Total		***101.58***	%
Additional parameters			
Insoluble residue	(3)	***81.2***	%
Fire loss	(2)	***4.22***	%
Bound water content	(2)	***2.6***	%
Organic content	(2)	***0.25***	%
CO_2 content	(2)	***1.11***	%
$CaCO_3$ equivalent content	(2)	***2.52***	%

PATHOLOGICAL PARAMETERS			
Parameters		Results	Units
Carbonation depth	(6)	***intl: 0 -ext: 13mm***	mm
Sulfate content	(7)	***0.49***	%
Free chlorides content	(3)	***< 0.02***	%
Chloride/cement content		***<0.15***	%
Chloride class		***Cl 0.20***	--
Soluble Na_2O content	(3)	***0.12***	%
Soluble K_2O content	(3)	***0.03***	%
Soluble Na_2O equivalent content		***0.14***	%
Soluble Na_2O equivalent content		***3.11***	**kg/m³**

NR: Not Reported Retention of samples: 1 month after the date of the analysis report

CODE	1	2	3	4	5	6	7	8	9	10	11
Internal method adapted from:	NF EN 12390-7	Bull LPC no. 5 JAN FEB 1961	NF EN 196-2	P 15-467	NF T90-007	NF EN 14630	MO MAT CCR 03	NF EN 1936	NF EN 12390-3	Pr EN 196-10	NF EN 15309

CONCRETEST® ANALYSIS REPORT

Date received:	20/05/2008	Site of sampling: NR
File reference:	12220/08/5646	
Sampling date:	NR	Sample reference: Coring no. 2
Sampling done by:	NR	Concrete age: NR

WEIGHT COMPOSITION OF CONCRETE		
Parameters	Results	Units
Cement content	11.4	%
Aggregates content	79.4	%
Silicate aggregates	90.3	%
Calcite aggregates	9.7	%
Water content	9.2	%
	100	%

CONSTITUTIONAL PARAMETERS		
Parameters	Results	Units
Cement content	265	kg/m 3
Water content	214	L
W/C ratio	0.81	--
Hydration rate	18	%
Entrained air content estimate	5.4	%

PHYSICAL PARAMETERS			
Parameters		Results	Units
Open porosity	(8)	9.5	%
Visibile density	(1)	2322	kg/m 3
Soaked density	(1)	2418	kg/m 3
Compressive strength	(9)	18.8	N/mm 3
Resistance class		C 16/20	--

CHEMICAL PARAMETERS			
Parameters		**Results**	**Units**
Basic parameters			
Fire loss 105°C / 450°C	(2)	***2.10***	%
Fire loss 450°C / 550°C	(2)	***0.22***	%
Fire loss 550°C / 950°C	(2)	***0.91***	%
Soluble silicate content	(5)	***0.19***	%
Na_2 content	(4)	***4.51***	%
MgO content	(4)	***1.68***	%
Al_2O_3 content	(4)	***10.9***	%
SiO_2 content (insoluble)	(4)	***55.9***	%
P_2O_5 content	(4)	***0.189***	%
SO_3 content	(4)	***1.63***	%
K_2O content	(4)	***2.18***	%
CaO content	(4)	***17.3***	%
TiO_2content	(4)	***0.201***	%
MnO content	(4)	***0.049***	%
Fe_2O_3 content	(4)	***1.75***	%
ZnO content	(4)	***0.013***	%
SrO content	(4)	***0.020***	%
PbO content	(4)	***0.012***	%
Total		***99.72***	%
Additional parameters			
Insoluble residue	(3)	***85.1***	%
Fire loss	(2)	***3.39***	%
Bound water content	(2)	***2.1***	%
Organic content	(2)	***0.22***	%
CO_2 content	(2)	***0.91***	%
$CaCO_3$ equivalent content	(2)	***2.07***	%

PATHOLOGICAL PARAMETERS			
Parameters		**Results**	**Units**
Carbonation depth	(6)	***intl: 0 -ext: 0mm***	mm
Sulfate content	(7)	***0.43***	%
Free chlorides content	(3)	***< 0.02***	%
Chloride/cement content		***<0.18***	%
Chloride class		***Cl 0.20***	--
Soluble Na_2O content	(3)	***0.07***	%
Soluble K_2O content	(3)	***0.03***	%
Soluble Na_2O equivalent content		***0.09***	%
Soluble Na_2O equivalent content		***2.09***	**kg/m^3**

NR: Not Reported   Retention of samples: 1 month after the date of the analysis report

CODE	**1**	**2**	**3**	**4**	**5**	**6**	**7**	**8**	**9**	**10**	**11**
Internal method adapted from:	NF EN 12390-7	Bull LPC no. 5 JAN FEB 1964	NF EN 196-2	P 15-467	NF 190-007	NF EN 14630	MO MAT CCR 03	NF EN 1936	NF EN 12390-3	Pr EN 196-10	NF EN 15309

CONCRETEST® ANALYSIS REPORT

Date received:	20/05/2008	Site of sampling: NR
File reference:	12220/08/5646	
Sampling date:	NR	Sample reference: Coring no. 3
Sampling done by:	NR	Concrete age: NR

WEIGHT COMPOSITION OF CONCRETE		
Parameters	Results	Units
Cement content	10.5	%
Aggregates content	79.1	%
Silicate aggregates	89.2	%
Calcite aggregates	10.8	%
Water content	10.4	%
	100	%

CONSTITUTIONAL PARAMETERS		
Parameters	Results	Units
Cement content	231	kg/m ^{3}
Water content	230	L
W/C ratio	1.00	--
Hydration rate	15	%
Entrained air content estimate	8.1	%

PHYSICAL PARAMETERS			
Parameters		Results	Units
Open porosity	(8)	14.9	%
Visibile density	(1)	2203	kg/m ^{3}
Soaked density	(1)	2352	kg/m ^{3}
Compressive strength	(9)	10.3	N/mm ^{3}
Resistance class		C 8/10	--

CHEMICAL PARAMETERS			
Parameters		**Results**	**Units**
Basic parameters			
Fire loss 105°C / 450°C	(2)	***1.57***	%
Fire loss 450°C / 550°C	(2)	***0.68***	%
Fire loss 550°C / 950°C	(2)	***1.98***	%
Soluble silicate content	(5)	***0.18***	%
Na_2 content	(4)	***3.49***	%
MgO content	(4)	***1.30***	%
Al_2O_3 content	(4)	***10.8***	%
SiO_2 content (insoluble)	(4)	***57.3***	%
P_2O_5 content	(4)	***0***	%
SO_3 content	(4)	***1.45***	%
K_2O content	(4)	***2.41***	%
CaO content	(4)	***16.5***	%
TiO_2 content	(4)	***0.225***	%
MnO content	(4)	***0.051***	%
Fe_2O_3 content	(4)	***1.74***	%
ZnO content	(4)	***0.012***	%
SrO content	(4)	***0.021***	%
PbO content	(4)	***0.018***	%
Total		***99.72***	%
Additional parameters			
Insoluble residue	(3)	***85.1***	%
Fire loss	(2)	***4.32***	%
Bound water content	(2)	***1.57***	%
Organic content	(2)	***0.68***	%
CO_2 content	(2)	***1.98***	%
$CaCO_3$ equivalent content	(2)	***4.49***	%

PATHOLOGICAL PARAMETERS			
Parameters		**Results**	**Units**
Carbonation depth	(6)	***intl: 0 -ext: 35mm***	mm
Sulfate content	(7)	***0.38***	%
Free chlorides content	(3)	***< 0.02***	%
Chloride/cement content		***<0.19***	%
Chloride class		***Cl 0.20***	--
Soluble Na_2O content	(3)	***0.09***	%
Soluble K_2O content	(3)	***0.03***	%
Soluble Na_2O equivalent content		***0.11***	%
Soluble Na_2O equivalent content		***2.42***	**kg/m^3**

NR: Not Reported Retention of samples: 1 month after tho date of the analysis report

CODE	1	2	3	4	5	6	7	8	9	10	11
Internal method adapted from:	NF EN 12390-7	Bull LPC no. 5 JAN FEB 1964	NF EN 196 2	P 15-467	NF T90-007	NF EN 14630	MO MAT CCR 03	NF EN 1936	NF EN 12390-3	Pr EN 196-10	NF EN 15309

CONCRETEST® ANALYSIS REPORT

Date received:	02/09/2008	Site of sampling: NR
File reference:	ZQ 7111	
Sampling date:	01/07/2008	Sample reference: /
Sampling done by:	NR	Concrete age: Around 50 years

WEIGHT COMPOSITION OF CONCRETE		
Parameters	Results	Units
Cement content	***11.2***	%
Aggregates content	***79.6***	%
Silicate aggregates	***97.2***	%
Calcite aggregates	***2.8***	%
Water content	***9.2***	%
	100	%

CONSTITUTIONAL PARAMETERS		
Parameters	Results	Units
Cement content	***254***	***kg/m³***
Water content	***208***	***L***
W/C ratio	***0.82***	--
Hydration rate	***19***	%
Entrained air content estimate	***5.1***	%

PHYSICAL PARAMETERS			
Parameters		Results	Units
Open porosity	(7)	***9.2***	%
Visibile density	(1)	***2261***	***kg/m³***
Soaked density	(1)	***2353***	***kg/m³***
Elution (length/diameter)		***1***	--
Compressive strength	(8)	***7.5***	***N/mm³***
Resistance class		***C 8/10***	--

CHEMICAL PARAMETERS			
Parameters		Results	Units
Basic parameters			
Fire loss 105°C / 450°C	(2)	***2.13***	%
Fire loss 450°C / 550°C	(2)	***0.22***	%
Fire loss 550°C / 950°C	(2)	***0.65***	%
Soluble silicate content	(5)	***2.95***	%
Na_2 content	(4)	***4.76***	%
MgO content	(4)	***1.59***	%
Al_2O_3 content	(4)	***9.49***	%
SiO_2 content (insoluble)	(4)	***48.58***	%
P_2O_5 content	(4)	***0.207***	%
SO_3 content	(4)	***1.44***	%
K_2O content	(4)	***1.75***	%
CaO content	(4)	***22.98***	%
TiO_2 content	(4)	***0.251***	%
MnO content	(4)	***0.046***	%
Fe_2O_3 content	(4)	***2.55***	%
ZnO content		***0.043***	%
SrO content	(4)	***0.023***	%
PbO content		***0.006***	%
Total		***99.66***	%
Additional parameters			
Insoluble residue	(3)	***84.9***	%
Fire loss	(2)	***3.2***	%
Bound water content	(2)	***2.13***	%
Organic content	(2)	***0.22***	%
CO_2 content	(2)	***0.65***	%
$CaCO_3$ equivalent content	(2)	***1.48***	%

PATHOLOGICAL PARAMETERS			
Parameters		Results	Units
Carbonation depth	(6)	***0***	mm
Sulfate content	(3)	***0.5***	%
Free chlorides content	(3)	***< 0.01***	%
Chloride/cement content		***<0.09***	%
Chloride class		***Cl 0.10***	--
Soluble Na_2O content	(3)	***0.07***	%
Soluble K_2O content	(3)	***0.04***	%
Soluble Na_2O equivalent content	(3)	***0.1***	%
Soluble Na_2O equivalent content		***2.26***	kg/m^3

NR: Not Reported Retention of samples: 1 month after the date of the analysis report

CODE	1	2	3	4	5	6	7	8	9	10	11
Internal method adapted from:	NF EN 12390-7	Bull LPC no. 5 JAN FEB 1964	NF EN 196-2	P 15-467	NF T90-007	NF EN 14630	Recomm-endations AFREM 1997	NF EN 12390-3	Pr EN 196-10	NF EN 15309	

A1.7.2. *Steel coating measurements.*

Quickscan: FQ000008.XFF

Date / Hour: 2009-06-11 10:32:25 Bar: 8mm SSN: 10308001

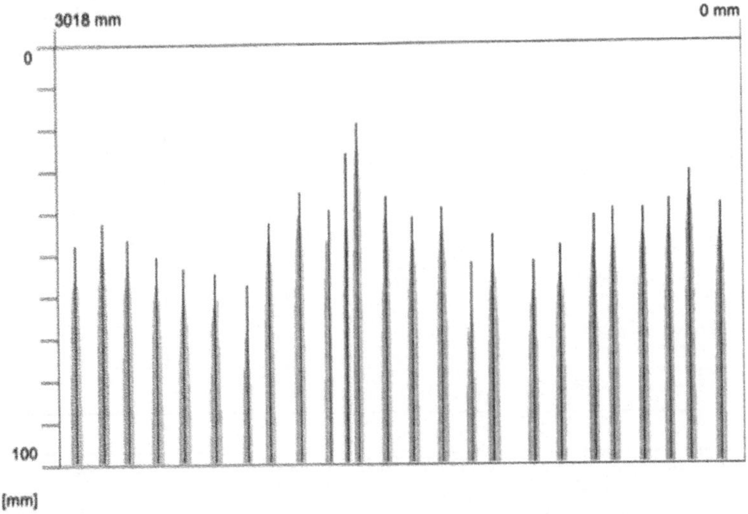

Quickscan Statistics:

Minimum depth:	19 mm	T1:	100 mm
Maximum depth:	58 mm	Number of bars above T1:	25
Average of bars:	42 mm	T2:	100 mm
Standard deviation:	9 mm	Number of bars above T2:	25
Cut-off:	100 mm	T3:	100 mm
Number of bars above cut-off:	25	Number of bars above T3:	25

Client: Syndicat

Location: Libourne Operator: XL

Comment:

Measurement of coating on the tank walls

Archiver le fichier: C:\Program Files\Hilti\PS200\5.3\Download\Prj00001\FQ000008.XFF

Quickscan: FQ000010.XFF

Date / Hour: 2009-06-11 10:37:44 Bar: 8mm SSN: 10308001

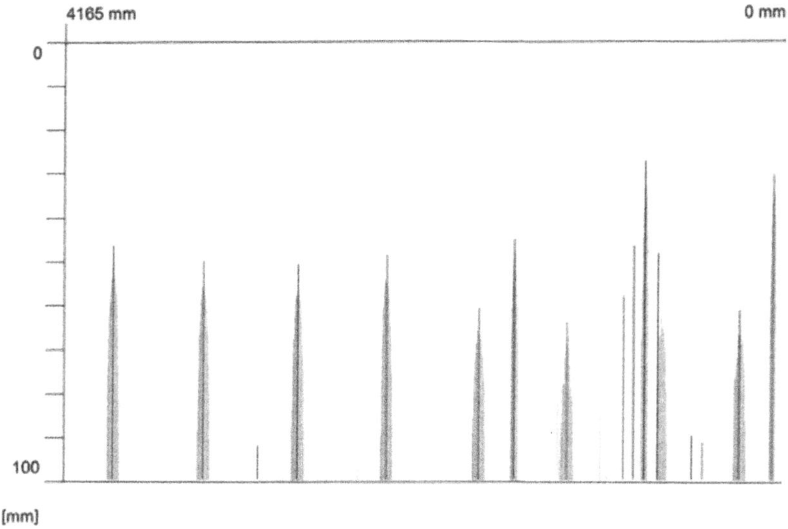

Quickscan Statistics:

Minimum depth:	27 mm	T1:	100 mm
Maximum depth:	92 mm	Number of bars above T1:	16
Average of bars:	56 mm	T2:	100 mm
Standard deviation:	20 mm	Number of bars above T2:	16
Cut-off:	100 mm	T3:	100 mm
Number of bars above cut-off:	16	Number of bars above T3:	16

Client: Syndicat

Location: Libourne Operator: XL

Comment:

Measurement of coating on the tank

Archiver le fichier: C:\Program Files\Hilti\PS200\5.3\Download\Prj00001\FQ000010.XFF

Quickscan: FQ000014.XFF

Date / Hour: 2009-06-11 11:19:36 Bar: 8mm SSN: 10308001

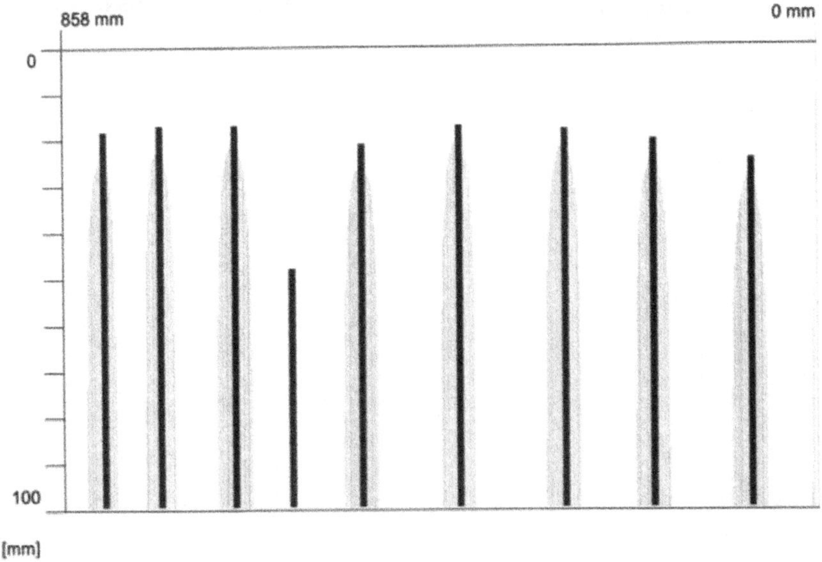

Quickscan Statistics:

Minimum depth:	17 mm	T1:	100 mm
Maximum depth:	48 mm	Number of bars above T1:	9
Average of bars:	22 mm	T2:	100 mm
Standard deviation:	10 mm	Number of bars above T2:	9
Cut-off:	100 mm	T3:	100 mm
Number of bars above cut-off:	9	Number of bars above T3:	9

Client: Syndicat

Location: Libourne Operator: XL

Comment:

Measurement of tank coating

Archiver le fichier: C:\Program Files\Hilti\PS200\5.3\Download\Prj00001\FQ000014.XFF

Quickscan: FQ000013.XFF

Date / Hour: 2009-06-11 11:00:53 Bar: 8mm SSN: 10308001

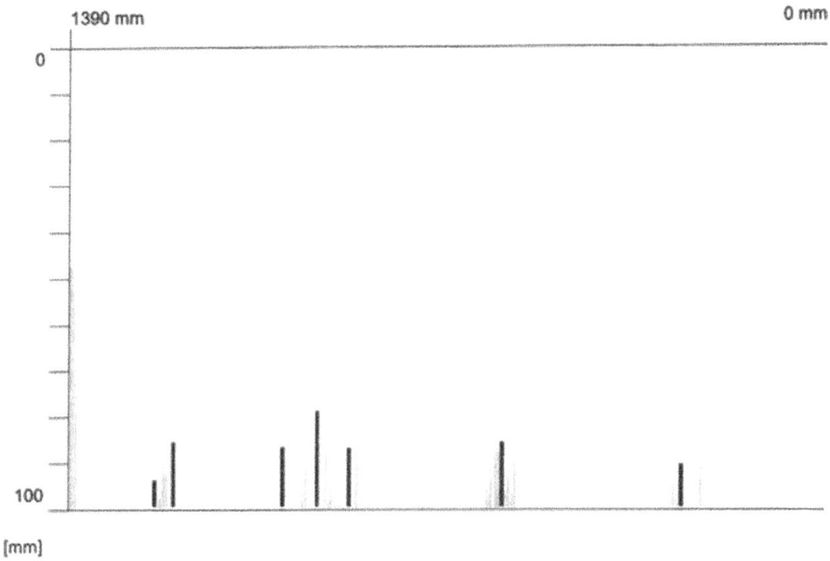

Quickscan Statistics:

Minimum depth:	79 mm	T1:	100 mm
Maximum depth:	94 mm	Number of bars above T1:	7
Average of bars:	87 mm	T2:	100 mm
Standard deviation:	5 mm	Number of bars above T2:	7
Cut off:	100 mm	T3:	100 mm
Number of bars above cut-off:	7	Number of bars above T3:	7

Client: Syndicat

Location: Libourne Operator: XL

Comment:

Measurement of tank dome coating

Archiver le fichier C:\Program Files\Hilti\PS200\5.3\Download\Prj00001\FQ000013.XFF

Quickscan: FQ000018.XFF

Date / Hour: 2009-06-11 11:19:36 Bar: 8mm SSN: 10308001

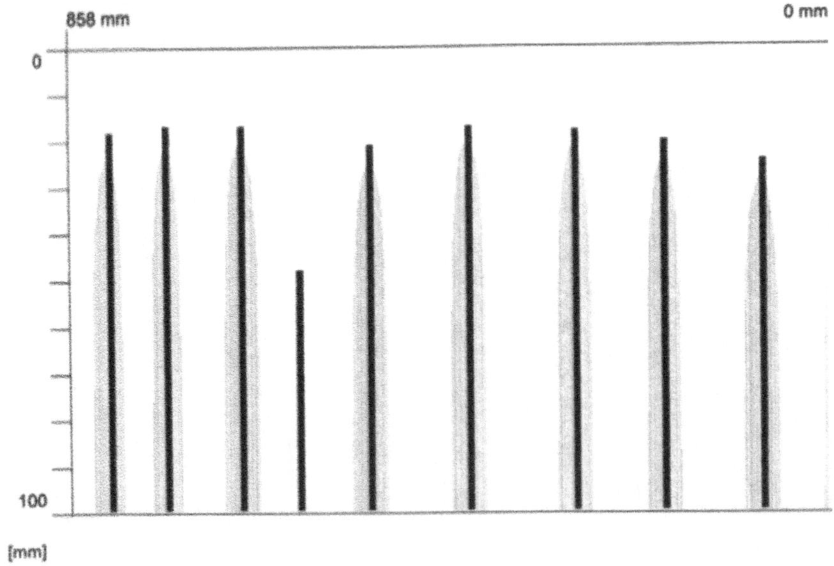

Quickscan Statistics:

Minimum depth:	17 mm	T1:	100 mm
Maximum depth:	48 mm	Number of bars above T1:	9
Average of bars:	22 mm	T2:	100 mm
Standard deviation:	10 mm	Number of bars above T2:	9
Cut-off:	100 mm	T3:	100 mm
Number of bars above cut-off:	9	Number of bars above T3:	9

Client: Syndicat

Location: Libourne Operator: XL

Comment:

Measurement of barrel wall coating

Archiver le fichier: C:\Program Files\Hilti\PS200\5.3\Download\Prj00001\FQ000018.XFF

Quickscan: FQ000020.XFF

Date / Hour: 2009-06-11 11:21:44 Bar: 8mm SSN: 10308001

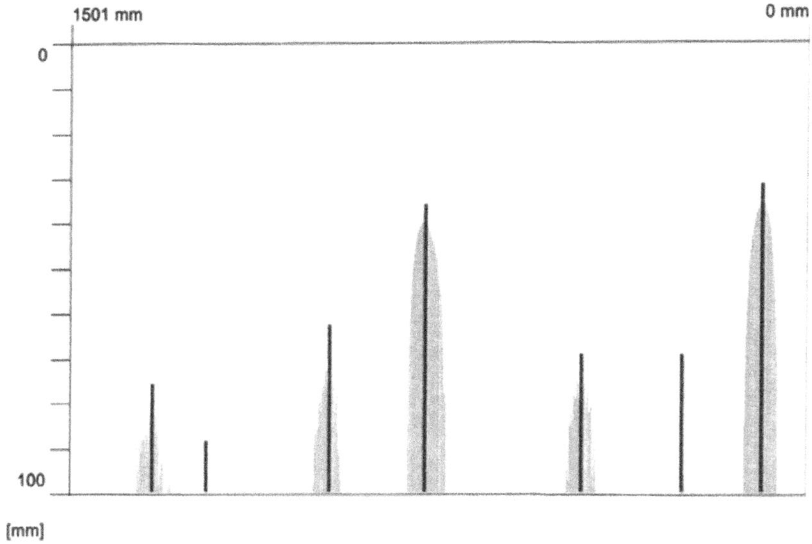

Quickscan Statistics:

Minimum depth:	32 mm	T1:	100 mm
Maximum depth:	89 mm	Number of bars above T1:	7
Average of bars:	62 mm	T2:	100 mm
Standard deviation:	21 mm	Number of bars above T2:	7
Cut-off:	100 mm	T3:	100 mm
Number of bars above cut-off:	7	Number of bars above T3:	7

Client: Syndicat

Location: Libourne Operator: XL

Comment:

Measurement of tank wall coating

Archiver le fichier: C:\Program Files\Hilti\PS200\5.3\Download\Prj00001\FQ000020.XFF

Quickscan: FQ000021.XFF

Date / Hour: 2009-06-11 11:23:16 Bar: 8mm SSN: 10308001

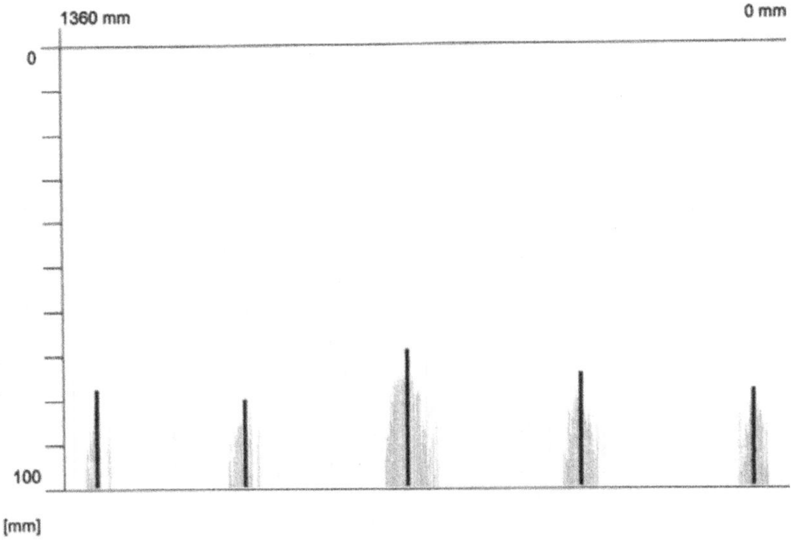

Quickscan Statistics:

Minimum depth:	69 mm	T1:	100 mm
Maximum depth:	80 mm	Number of bars above T1:	5
Average of bars:	75 mm	T2:	100 mm
Standard deviation:	4 mm	Number of bars above T2:	5
Cut-off:	100 mm	T3:	100 mm
Number of bars above cut-off:	5	Number of bars above T3:	5

Client: Syndicat

Location: Libourne Operator: XL

Comment:

Value of the tank wall coating

Archiver le fichier: D:\MESDOC~1\FQ000021.XFF

A1.7.3. *Photographs of the structure*

	Photo 1 General view in elevation
	Photo 2 View of column sampling

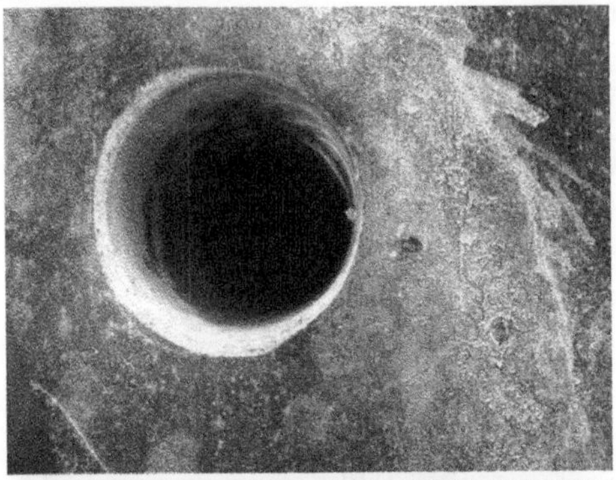	Photo 3 Column sampling
	Photo 4 Beginning of corrosion of the reinforcements after carbonation of the concrete

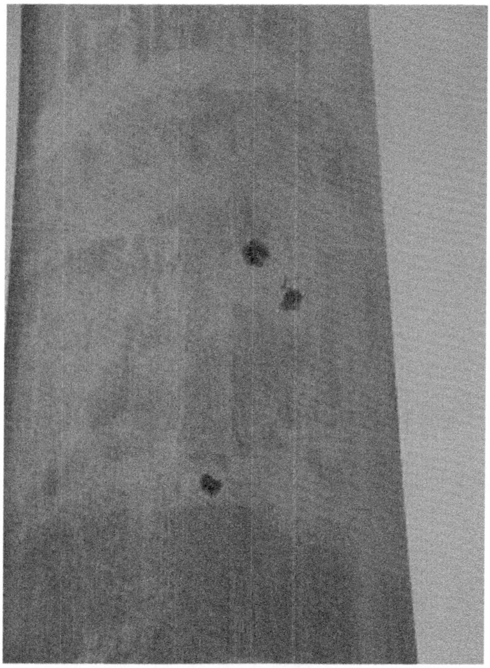	Photo 5 Beginning of corrosion of the reinforcements after carbonation of the concrete
	Photo 6 Crazing of the coating

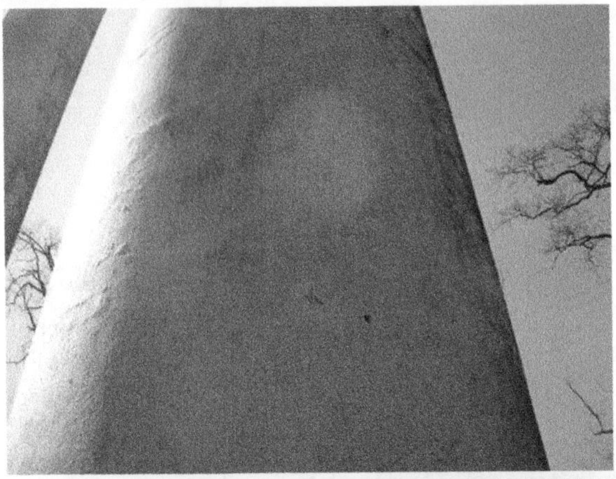

Photo 7
Overview of corrosion bursts of reinforcements behind the coating

Photo 8
Deterioration of column paint

	Photo 9 Crazing of the coating. Cracking of the recovery casting. Cracking of the coating in the rebound blast area.
	Photo 10 Crazing of the tank coating and initiation of cracking in the rebound blast area (above the cornice)
	Photo 11 Crazing of the tank coating and initiation of cracking in the rebound blast area (above the cornice)

Appendix 2

Examples of Diagnosis on a Petroleum Products Storage Tank According to the DT 92 Method

A2.1. Origin and extent of the mission

According to our order no. 4501313058 on November 6, 2012, we were entrusted with a detailed inspection mission of the concrete container forming a retention tank for the storage of ammonia R5570.

With our operating authorization providing for a monitoring plan for this storage, we have incorporated a control plan for the concrete container.

In May 2011, a monitoring guide for civil engineering structures and UIC-UFIP structures (DT92) was released in order to address plans to modernize French industrial facilities. We wish to integrate this guide into our control plan. Our detailed inspection is based on this guide.

A2.2. Description of the structure

The procedures for monitoring the aging of structures depend on how hazardous the products stored within are. The structures are classified into two categories:

– structures listed in the modernization plan that are not in category II are classified in category I;

– the storage of flammable liquids and the "most critical" structures as defined by the professional guide for the definition of a perimeter as part of a modernization plan (notably flammable liquid containment vessels targeted by the ministerial decree 1432 and retention tanks for liquids with risk phrases R50 (including ammonia) and R50/53 for those greater than 100 m^3).

If the storage of ammonia R5570 exceeds 100 m^3, the retention tank is among the "most critical" structures and is therefore classified in category II. This 10,000 T storage tank consists of a heat-resistant metal tank that is 28 m in diameter and 26 m high.

This tank is placed inside a concrete structure consisting of a 50 cm thick foundation supported by piles and a cylindrical container made up of prestressed concrete. This prestressed concrete container, subject to inspection, has an internal diameter of 31 m, a height of 25.10 m and a thickness of 50 cm.

It has a low belt of 50 cm at the level of the concrete apron, a high belt that is 90 cm thick and 1 m high, and three vertical ribs for tensioning the prestressed cables. The construction dates back to 1996. Date of last inspection visit: not applicable. The concrete container has never been diagnosed: this is the first inspection carried out by an external company. Repairs carried out since the last monitoring action: not applicable.

A2.3. Investigation method

Investigations took place on November 26, 2012 in the presence of Mr. ... inspector in charge of ....

Weather conditions: rain, 8 °C.

Safety measures carried out before intervention: *degassing of the inner annular space.*

We began by inspecting the inside of the container, in which we were able to locate ourselves because of the numbering in place. We then inspected the *exterior*: the three vertical ribs delimit *three sectors that we examined using a 24 m telescopic aerial basket* mounted on a tractor with extensible axles as well as using *the service staircase for the area that was not accessible by aerial basket* (southern area).

A2.4. Nature of damages and explanations

The damages we identified are described below and illustrated by the photos. Notation of damages is presented in Table A2.1.

Condition class of structures	Definition of class	Comments	Nature of the intervention
1	Satisfactory condition only requiring routine maintenance		Cleaning of tanks and splitting joints Cleaning of drainage Control of tank access devices, pipes, etc.
2	Fair condition with mild damage beyond routine maintenance	Specialized maintenance should be provided	Drainage repairs Recovery of splitting joints Repair of local damage (small cracks, spalling, etc.) Treatment of corrosion of metallic elements Repair of sealants and fire protection provisions.
2E	Ditto state 2 but with risk of evolution of pathologies (evolutionary state)	Implementation of reinforced control	
3	Degraded structural condition requiring repair work	Diagnosis and repair	Major structural repairs (walls, paving, foundations, etc.) Replacing anchor bolts Implementation of instrumentation of the structure
3P	Ditto state 3 but with a priority deadline (integrity, retention capacity, bearing capacity that can be faulted quickly)	Diagnosis and repair as soon as possible	

Table A2.1. *Notation of damages*

A2.4.1. *Vertical ribs*

A2.4.1.1. *Southeast rib*

Southeast rib		(observation sheet no. 001)
Findings	Level	Photos
Multiple horizontal cracks and microcracks <0.2 mm depending on the reinforcement in the *highest ring*. This defect is due to *lack of sufficient coating* of the steels.		

A2.4.1.2. *Northeast rib*

Northeast rib		(observation sheet no. 002)
Findings	Level	Photos
Multiple horizontal cracks and microcracks <0.2mm depending on the steel reinforcement in the *highest ring*.		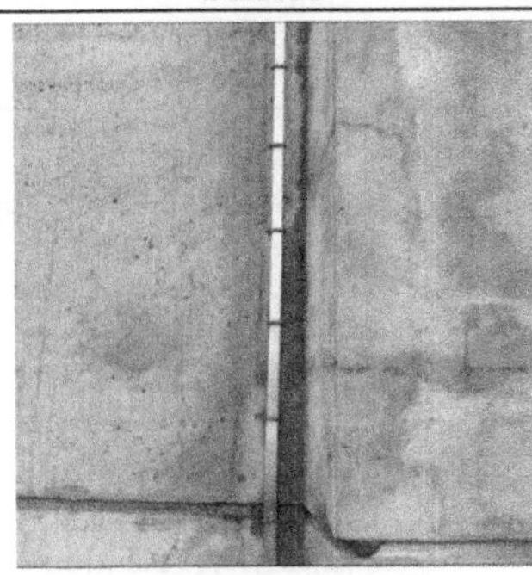
Shards of concrete with visible reinforcements that are altered very little at the junction between *the rib and the ring*		

A2.4.1.3. *West rib*

West rib		(observation sheet no. 003)
Findings	Level	Photos
Shards of concrete with visible reinforcements that are altered very little at the ridge. This damage is due to the oxidation of reinforcements, which caused a push on the concrete. This is the evolution of the damage described just before, namely cracking from a coating defect.		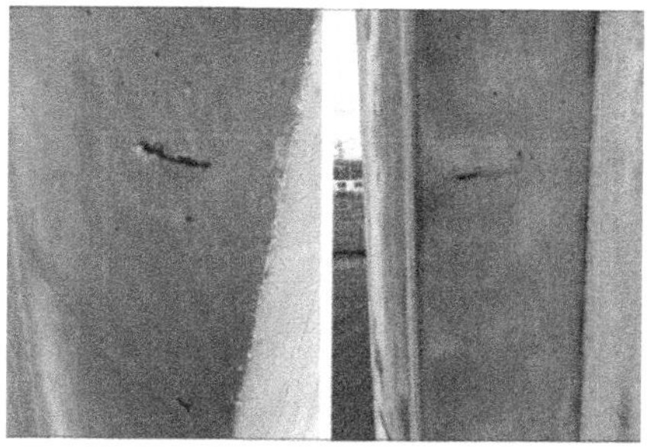

A2.4.2. *Rings*

A2.4.2.1. *East façade*

East façade		(observation sheet no. 004)
Findings	Level	Photos
Multiple cracks and microcracks that are mainly horizontal <0.2mm in the prestressed concrete skirt. This damage is certainly due to a coating *defect of the passive steel reinforcement* and retraction of the concrete.		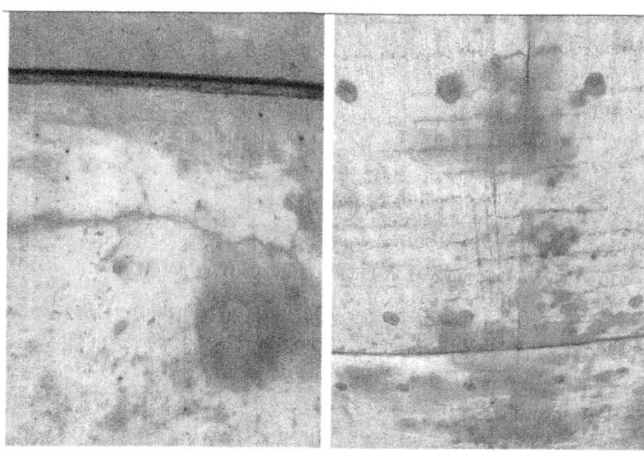

Findings	Level	Photos	
Surface shrinkage of concrete due to shrinkage on the prestressed concrete skirt of the *highest ring*. This damage is due to overly fast desiccation.	1	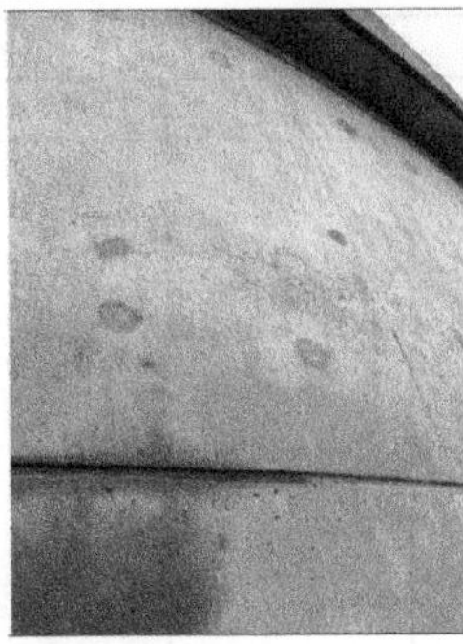	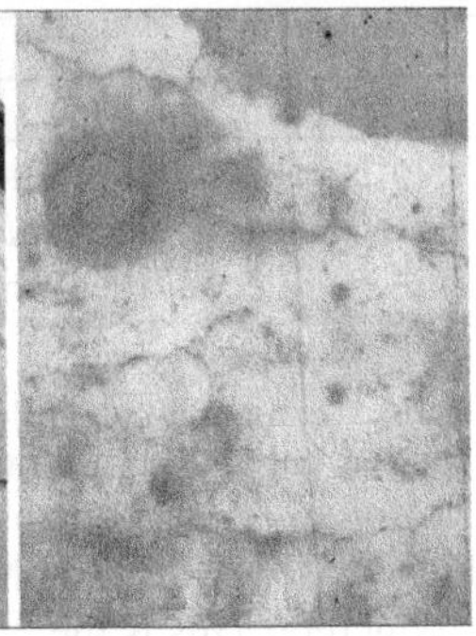

A2.4.2.2. *North-West façade*

North-West façade		(observation sheet no. 005)	
Findings	Level	Photos	
Surface shrinkage of the concrete due to shrinkage at the junction between the rib and the prestressed concrete skirt of the last and the penultimate rings starting from the top. This damage is due to the overly fast desiccation of concrete (lack of curing at the implementation).	1		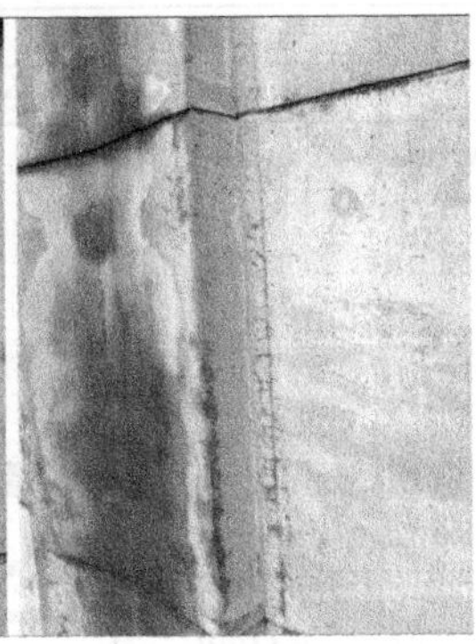

A2.4.2.3. *Space inside the ring*

Space inside the ring		(observation sheet no. 006)
Findings	**Level**	**Photos**
Horizontal cracks and microcracks <0.2 mm: – at the 60H marker at 2 m height over a length of 60 cm; – between the 60H and 64H markers at 1.50 m height over a length of 50 cm.	1	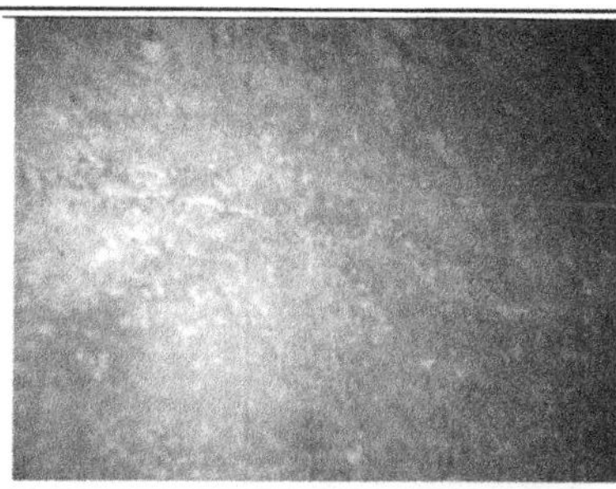 
Surface shrinkage of concrete due to shrinkage on the prestressed concrete skirt between 20H and 24H markers that are between 1.5 and 2.5 m. This damage is due to the overly rapid desiccation of concrete (lack of curing at the *implementation*).	1	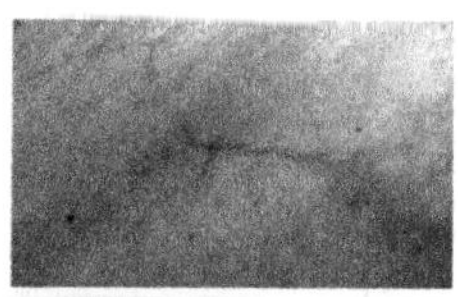 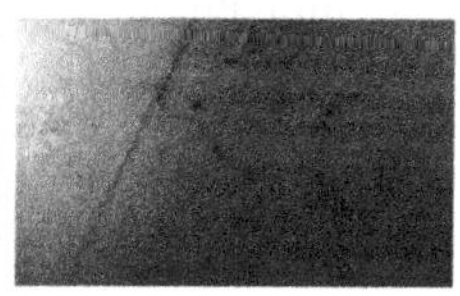 

A2.4.3. *Concrete apron*

A2.4.3.1. *East façade*

East façade		(observation sheet no. 007)
Findings	Level	Photos
Concrete shards with visible reinforcement with little alteration at the junction with the apron (presence of a ball joint per section of Freyssinet-type shrunken concrete). This damage is initially due to the rotation of the skirt with respect to the foundation creating the crack and, in a second stage due to the oxidation of reinforcements causing a push on the concrete.	2E	 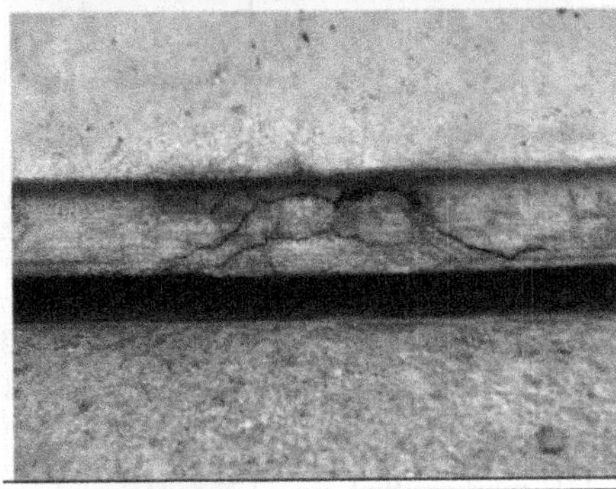

A2.5. Executive summary of the condition of the structure and its evolution

Reminder of conclusions from the last management actions of the structure	Not applicable: this is the first detailed inspection.
Interpretations of measures and recognition	Not applicable.
Analysis of the causes and significance of the damage	We were able to identify two different levels of damage: *Level D1 damage*: – On the outside of the concrete container, multiple

	horizontal cracks and microcracks <0.2 mm on the reinforcement in the vertical South-East rib (highest ring), North-East rib (third ring from the bottom), Western rib over almost the entire height and rings on the East side. This damage is likely due to inadequate coating of the reinforcement. – On the outside of the concrete container, superficial surface cracking due to shrinkage on the prestressed concrete skirt on the East façade on the highest ring, on the North-West façade on the last and penultimate rings from the top. This damage is due to the overly rapid desiccation of concrete (lack of curing in the implementation phase). Inside the concrete container, we saw the same damage between the 20H and 24H markers that are between 1.5 and 2.5 m. – Inside the concrete container, horizontal cracks and microcracks <0.2 mm at the 60H marker at <2 m height over a length of 60 cm, between the 60H and 64H markers at 1.50 m over a length of 50 cm (likely appeared during construction and, in our view, was not evolutionary). *Level D2E damage*: – On the outside of the concrete container, concrete fragments with visible reinforcements that have had very little alteration, mainly on the West and North-East vertical ribs. This damage is due to oxidation of reinforcements. – This damage is found at the junction with the apron on the East façade. This damage was initially due to the rotation of the skirt relative to the foundation thus creating a crack and, in a second stage, oxidation of the reinforcements leading to a push on the concrete.
Opinion on the condition of the structure	In light of the damages observed, the structure should be considered as class 2E: presence of damages that risk evolving reinforced control to be considered. The damages observed are superficial and do not affect the strength of the structure or its end-use. Evolution of the damages should be monitored.
Suggestions for maintenance and repairs: – Systematic routine maintenance – Specialized maintenance – Repairs/confirmation in order of priority	Only routine maintenance is required.

Suggestions for development	To inspect the inner annular space, provisional lighting should be improved. To avoid stagnation of water in the annular space, which will eventually corrode the steels in the mounting cap at the bottom of the skirt, waterproofing in the dome needs to be improved.
Proposals for specific investigations or monitoring	Reinforced control will need to be set up at the skirt/apron junction in order to know the evolutive nature of the damages observed, bearing in mind the mechanical operation of this part of the structure (section of shrunken concrete = joint) and the presence of water in the annular space.
Proposals for security measures	A crinoline and resting bearings should be fitted between the various elements of the existing ladder in the inner annular space.
Proposed amendments to the monitoring routine	The DT92 guide requires monitoring visits at an annual frequency for category II structures, despite the general condition of the structure.

Appendix 3

Examples of Diagnosis of a Marine Structure Using the CETMEF VSC Method

A3.1. Appendix 3a: Periodic detailed inspection of 2009 campaign

A3.1.1. *Structure: pier. Case study*

Figure A3.1. *General view*

A3.1.1.1. *Description of the structure*

A3.1.1.1.1. Location

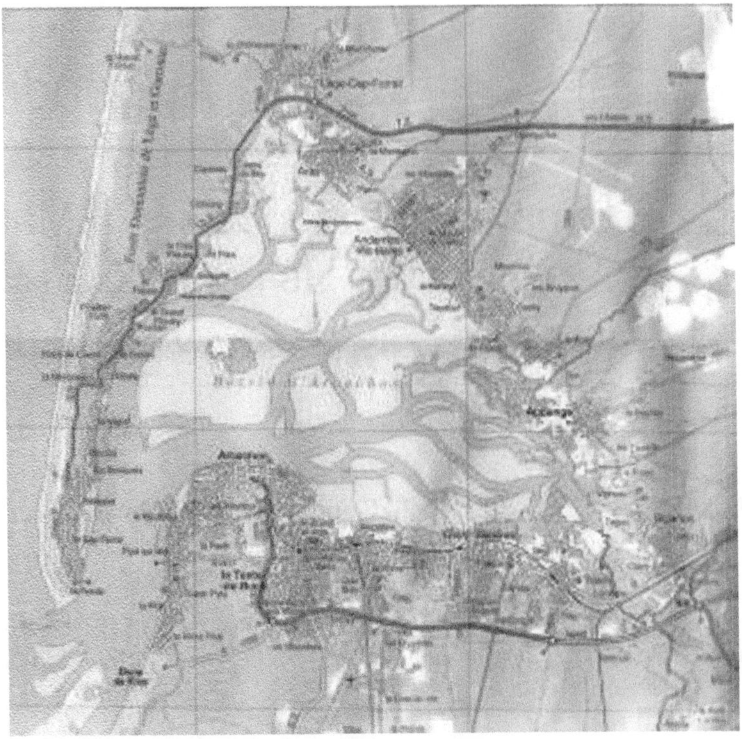

Figure A3.2. *Aerial view of site*

A3.1.1.1.2. Identification of the structure

Name of the structure:	Andernos pier
Identification number:	Not disclosed
Town:	Andernos
Structure class:	Not applicable

A3.1.1.1.3. General description of the structure

Nature of the structure

It is a concrete pier constructed in two parts:

– the first part consists of reinforced concrete columns on which precast concrete beams rest;

- the columns have a diameter of 40 cm and they support beams that are 25 cm wide at 50 cm height. A reinforced concrete slab that is 4 m in length and 0.15 m thick is supported by the two rows of beams. This portion of the pier was renovated in 1994. It is not part of this inspection,

– the second part is constructed from reinforced concrete elements: circular columns that support rectangular beams and a prefabricated slab.

The characteristics are summarized in Table A3.1.

Foundations	Not visible
Columns	Reinforced concrete with diameter of 600 mm
Beams	Reinforced concrete beams 0.35 m wide and 0.50 m high
Slabs	Prefabricated reinforced concrete slabs 0.18 m thick
Equipment or superstructures	Aluminum railing

Table A3.1. *Characteristics of the structure*

A3.1.1.1.4. Dimensional characteristics

Total length of structure	152 m + 77 m = 229 m
Length of the inspected part	77 m
Column spacing perpendicular to the direction of movement	3 m
Column spacing parallel to the direction of movement	Around 7 m
Width of the slab	4 m

A3.1.1.1.5. Conditions for use

It is a promenade pier designed exclusively for pedestrians or cycling.

A3.1.1.2. *Conditions for implementation of the structure*

A3.1.1.2.1. Construction of the structure

Company	Unknown
Consultancy	Unknown
Years of construction	Unknown

A3.1.1.2.2. Reported incidents

Falling of concrete blocks from the underside of the pier.

A3.1.1.2.3. Organized monitoring of the structure

None.

A3.1.1.2.4. Work done since last visit

Purge and recovery of degraded elements are currently under way by Chantiers d'Aquitaine.

A3.1.1.3. *Inspection information*

Organization responsible for this operation	
Previous inspection	None
Date of this operation	May/June 2009
Breakdown of the inspection team	One team consisting of one civil engineer and one general engineer
Other participants	Core drilling company
Means used	Visual examination Ferroscan tests Destructive testing Sclerometric tests
Atmospheric conditions	Variable time
Other conditions	Tides
Particular difficulty	None

A3.1.1.4. *Inspection of the structure*

A3.1.1.4.1. Identification

The pier is located in a maritime area at the bottom of the Arcachon basin.

Given the tidal activity of the Arcachon basin, the columns are alternately located in the submerged area and in open air.

A3.1.1.4.2. Access to the structure

Elements	Description	Location	Opinion	Photo no.
Access to the structure	A structure located within a maritime area that is accessible from an urban area	–	Satisfactory condition	–
Roadway	Paved way occasionally accessible to vehicles	–	Satisfactory condition	–

A3.1.1.4.3. Superstructures

Reminder on the CETMEF's VSC categorization

Mechanical state index EIm:

EIm index	Evaluation of the condition
1	Structures with severe mechanical damage and immediate risk of ruin
2	Structures presenting severe mechanical damage without risk of immediate ruin
3	Structures with minor degradation or pathology
4	Structure in good condition

Status indicator EIu:

EIu index	Evaluation of the condition
1	Elements presenting degradations that may generate immediate safety problems
2	Elements presenting degradations that are likely to generate operating problems
3	Elements presenting degradations that are likely to generate problems of discomfort
4	Elements in good condition

The status index of structures is then defined as: EI = Min (EIm; EIu).

Pier support pillars

– Visible damage in structures:

- the visible defects are listed in Appendix A3.2;
- they mainly consist of vertical cracks in the concrete coating area.

There was no evidence of any pathology in the characteristics of a structural defect of these elements.

– State of conservation of structures:

The apparent state of conservation of the columns can be considered as good.

No corrosion was detected in the reinforcements because of substantial coating of the steels (see the pachometric test results in the Appendix).

Inspections were carried out using the Ferroscan PS 200 for verifying the reinforcements on the external faces of the columns, and point measurements for the mechanical characteristics of concretes were carried out by sclerometer.

The values found are summarized in Table A3.2.

Location	Inspection method	Measured covering (mm)		Regulatory covering
Column	Horizontal scanning Vertical scanning	Min: 44 Min: 46	Max: 87 Max: 98	Min: 50 mm

Table A3.2. *Measured covering*

The values of the full tests are in the Appendix.

Laboratory tests brought to light the following:

Page 1 of 2

CONCRETEST® ANALYSIS REPORT

Date received: 08/06/2009 Sampling site : NR
File reference number: Andernos-les-Bains jetty - tidal zone
Sampling date: NR Sample reference : Coring on 2 June 2009
Sampling done by: NR Concrete age : NR

WEIGHT COMPOSITION OF CONCRETE		
Parameters	Results	Units
Cement content	*13.8*	%
Aggregates content	*75.2*	%
silicate aggregates	*94.2*	%
calcite aggregates	*5.8*	%
Water content	*11.1*	%
Total	*100*	%

CONSTITUTIONAL PARAMETERS		
Parameters	Results	Units
Cement content	***301***	**kg/m³**
Water content	***242***	***L***
W/C ratio	***0.80***	**–**
Hydration rate	***18***	**%**
Entrained air content estimate	***8.4***	**%**

PHYSICAL PARAMETERS			
Parameters		Results	Units
Open porosity	(6)	***15.4***	%
Visible density	(6)	***2187***	kg/m³
Soaked density	(6)	***2341***	kg/m³
Elution (Length/Diameter)		**2**	--
Compressive strength	(7)	***NR***	***N/mm²***
Resistance class		***NR***	--

CHEMICAL PARAMETERS			
Parameters		Results	Units
Basic parameters			
Fire loss 105°C / 450°C	(1)	2.42	%
Fire loss 450°C / 550°C	(1)	0.28	%
Fire loss 550°C / 950°C	(1)	2.09	%
Soluble silicate content	(4)	3.91	%
Na_2 content	(3)	2.66	%
MgO content	(3)	1.57	%
Al_2O_3 content	(3)	6.90	%
SiO_2 content (insoluble)	(3)	47.2	%
P_2O_5 content	(3)	0.213	%
SO_3 content	(3)	2.18	%
K_2O content	(3)	1.36	%
CaO content	(3)	25.4	%
TiO_2 content	(3)	0.221	%
MnO content	(3)	0.067	%
Fe_2O_3 content	(3)	1.58	%
ZnO content		0.022	%
SrO content	(3)	0.022	%
PbO content	(3)	0.01	%
Total		98.11	%
Additional parameters			
Insoluble residue	(2)	79.1	%
Fire loss	(1)	5.04	%
Bound water content	(1)	2.42	%
Organic content	(1)	0.28	%
CO_2 content	(1)	2.09	%
$CaCO_3$ equivalent content	(1)	4.74	%

ADDITIONAL PARAMETERS - Free chlorides 3		
Sampling site	**Results**	**Units**
Surface	1.14	%
Middle: 5cm	0.52	%
Reinforcement: 10cm	0.48	%

INTERPRETATION OF RESULTS.–

The above results allow us to highlight the following elements:

– the composition of the concrete includes a mixture of cement and siliceous aggregates (about 94% silicate elements). The cement dosage is relatively low, around 300 kg/m^3 for a water content of 242 L/m^3, so a W/C ratio of around 0.80. The standard for EN 206.1 concrete would classify this environment as XS3 (tidal zone). This classification leads to an equivalent binder dosage of 350 kg/m^3 for a W/C ratio of 0.50 and a minimum compressive strength of 35 MPa. As a result, the concrete in place is largely underdosed for cement and overdosed with water, resulting in a concrete that is not very compact and is very sensitive to penetration of sea water;

– determining the hydration rate gives a value of about 18% for an expected value of 17%, which validates normal setting of the binder in the poured concrete;

– the concrete porosity is measured in at 15.4% for a density of 2,187 kg/m^3. These values corroborate the previous remark on the production of a relatively porous concrete;

– the dosage determination of free chlorides (which are soluble in sea water) was carried out at three different depths: at the surface (sea side), at –5 cm and at –10 cm from the surface. The values obtained are significant since they are 1.14% at the surface, 0.52% at –5 cm and 0.48% at –10 cm. The average chloride dosage across the entire coring is about 0.63%; at the reinforcement located at –10 cm from the surface, the chloride content relative to the cement dosage is 3.5%. Corrosion of reinforcements by chlorides can be seen in the sample; however, it should be noted that it is at –10 cm from the surface (the regulatory coating for a class XS3 concrete is 5 cm).

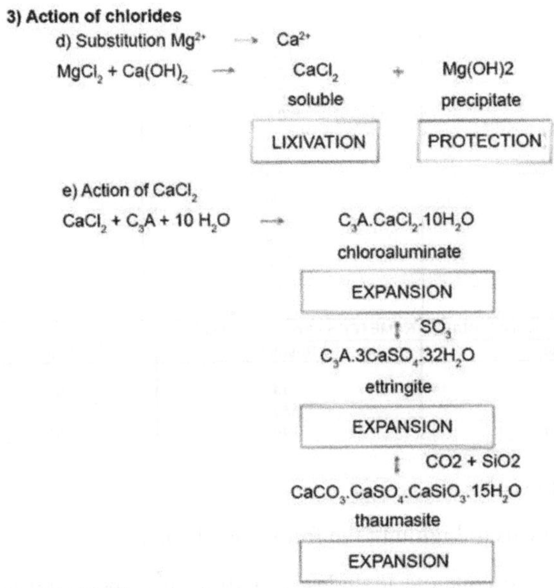

Figure A3.3. *Action of chlorides on a concrete structure*

PRINCIPLE OF ATTACK OF CHLORIDES.–

– the alkaline content in the analyzed concrete is 0.19% and is above the limit values recommended by the LCPC (4.16 kg/m³ for a recommended value of 3.5 kg/m³). This value is probably linked to the sodium dosage from seawater rather than to the risk assessment of alkali reaction. However, this phenomenon, which can change over the long term, will have to be monitored;

– the measurement of carbonation is zero.

The beams in the pier

– Visible damage in structures:

- the beams have many defects:

These defects are the central topic of an exhaustive statement attached in the appendix.

In summary, there are concrete cracks that occurred due to corrosive stresses of reinforced concrete steels.

The latter are sometimes absent, making the pathology structural.

– Condition of structures:

- visual and instrumental inspection:

Inspections were carried out on the Ferroscan PS 200 to verify the reinforcements of the external faces and the underside of the beams. A sclerometer was also used to take point measurements of the mechanical characteristics of concretes.

The values found are summarized in Table A3.3.

Location	Inspection method	Measured covering (mm)		Regulatory covering
Side view	Ferroscan	Min: 14	Max: 100	Min: 50 mm
Underside	Ferroscan	Min: 20	Max: 90	Min: 50 mm

The complete digital outputs of the measurements are in the Appendix.

Table A3.3. *Measured covering*

The slabs in the pier

– Visible damage in structures:

- the slabs have no visible structural defect:

Some mechanical damage to surface concretes has been noted, but these do not reveal major structural pathologies.

– Condition of structures:

- visual and instrumental inspection:

Ferroscan PS 200 inspections were carried out to verify the coatings of the undersides of the slabs. A sclerometer was also used to take point measurements of the mechanical characteristics of the concretes.

The values found are summarized in Table A3.4.

Location	Inspection method	Measured covering (mm)	Regulatory covering
Underside	Ferroscan	Min: 33 Max: 61	Min: 50 mm

Table A3.4. *Measured covering*

NOTE.– Due to their visible condition, laboratory tests were not done on beams and slabs, but the results obtained for columns are also valid for these elements.

A3.1.1.5. *Summary*

A3.1.1.5.1. Defects found

Defects	Cause	Remedies
	Construction defects	
Beams: bursting of concrete in beams	Poor compactness of concrete, insufficient covering	Significant repair or replacement
Slab	Insufficient covering	Protection or replacement
Columns	Locally insufficient covering	Protection
	Lack of maintenance defects	
Not applicable		
	Operational defects	
Not applicable		

A3.1.1.5.2. Degree of corrosion

Aluminum railing without any noticeable corrosion.

A3.1.1.5.3. Analysis of operating conditions

Not observed.

A3.1.1.5.4. Operation of existing measuring devices

Not applicable.

A3.1.1.5.5. Interpretation of observations

Not applicable.

A3.1.1.5.6. Evolution compared to the previous detailed inspection

No inspection provided.

A3.1.1.6. *Conclusion*

A3.1.1.6.1. State of the structure

Parts of the structure	Probable causes	Mechanical status index EIm	Status index EIu	Possible repair mode
Pier support pillers	No pathology despite localized coating defect	3		Cathodic protection
Beams	Generalized defect in coatings	1		Replacement and cathodic protection
Slabs	No pathology despite localized coating defect	3		Cathodic protection or replacement
Railing	No pathology despite localized coating defect		4	
Overall structure (maximum severity index)		1		

Table A3.5. *State of structure*

A3.1.1.6.2. Proposed work

Routine maintenance

None.

Specialized maintenance

None.

Major repairs

See Table A3.5.

Modernization

None.

A3.1.1.6.3. Further investigations to be undertaken

None.

A3.1.1.6.4. Estimation of the reinforcement and repair work to be considered

None.

A3.2. Appendix 3b: Directory of pathologies

Beam C-12/13	Defect no. 1
Healthy outer face except near the support line 13: slight crack in the top part of the beam over 30 cm.	
	No sign of corrosion

Beam C-12/13	Defect no. 2
Inside face and lower face 1.35 m from line 13. Longitudinal crack 70 cm long at 6 cm from the corner with the underside of the beam. Corresponding longitudinal crack in the underside of the beam at 13 cm from the corner.	

	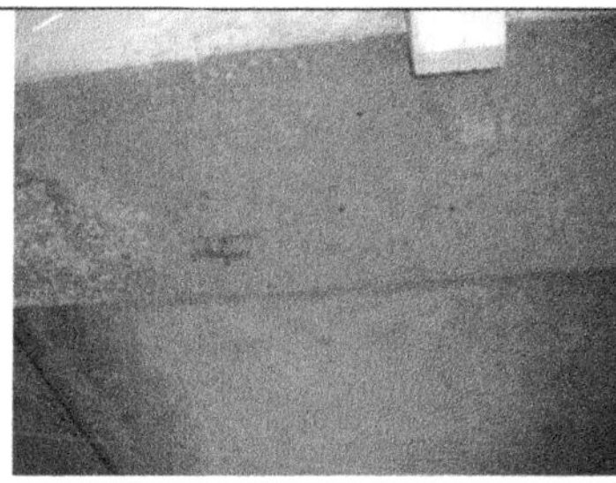
	Slight traces of corrosion

Beam C-12/13	Defect no. 3
Inside face 3.20 m from line 12. Longitudinal crack 60 cm long at 6 cm from the lower corner.	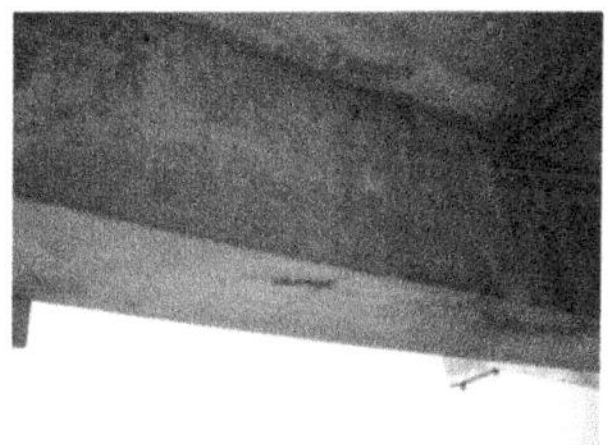
	No sign of corrosion

Beam C-12/13	Defect no. 4
Inner side near line 13. Longitudinal crack in the top part of beam 8 cm from the corner.	
	Slight trace of rust

Beam C-12/13	Defect no. 5
Underside. Longitudinal crack. Some mid-range transverse cracks.	
	No sign of rust

Beam C-11/12	Defect no. 1
Outside. Three longitudinal cracks: From support 12 across a length of 1.30 m, trace of rust at 4 cm from the underside. At 2.50 m from support 12 over 0.20 m. At 1.80 m from support 11 up to support 11: crack with a 5–10 mm opening.	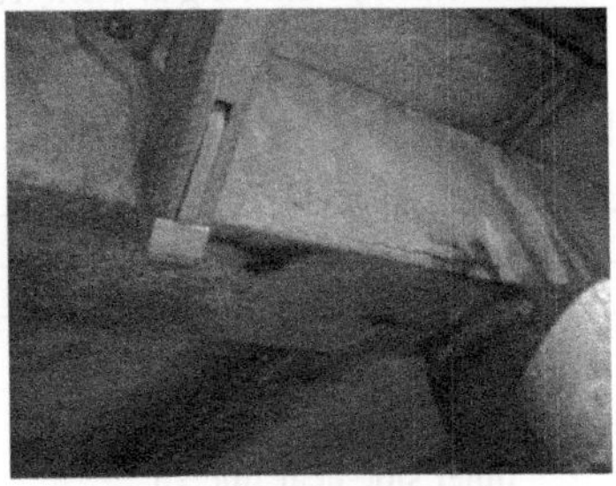

Beam C-11/12	Defect no. 2
Inside. Over 0.50 m from support 12 at 3 cm from the underside. Over 0.60 m at 3 cm. Over 0.70 m at 0.80 m from support 11.	

Beam C-11/12	Defect no. 3
Underside. Traces of corrosion.	

Beam C-10/11	Defect no. 1
Outside. Nothing to report.	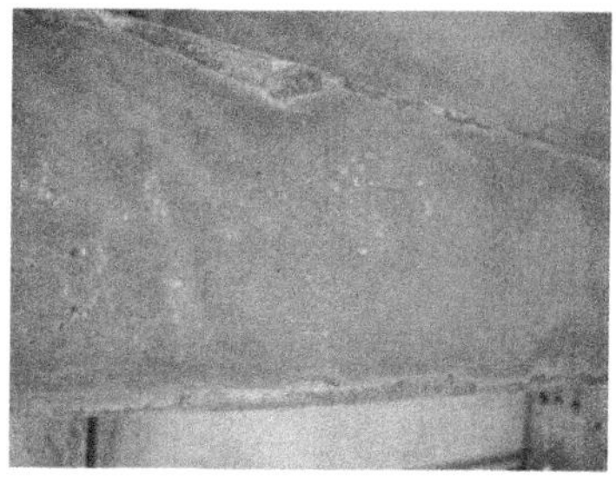

Beam C-10/11	Defect no. 2
Inside. One crack of 0.30 m at 3 m from support 11.	

Beam C-10/11	Defect no. 3
Underside. Nothing to report.	

Beam C-9/10	Defect no. 1
Outside. Three longitudinal cracks: Length of 1.60 m starting at 0.07 m from the corner. Length of 0.30 m at 0.04 m from the corner. Length of 2.90 m at 0.07m from the corner.	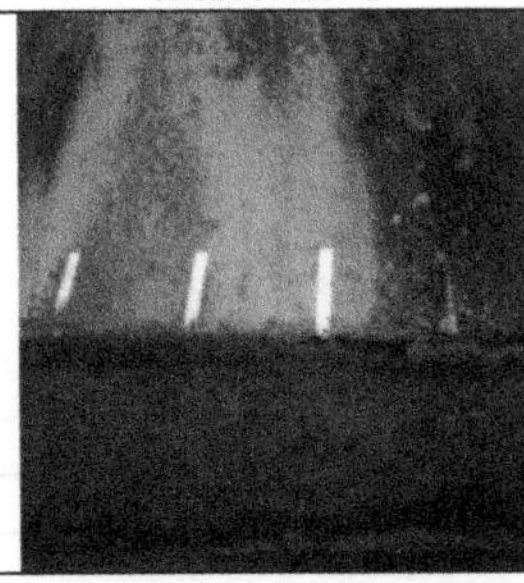

Beam C-9/10	Defect no. 2
Inside. One crack along the length of the beam at 0.08 m from the corner at the bottom. One longitudinal crack in the upper part over the entire length starting at 1 m from support 9.	

Beam C-9/10	Defect no. 3
Underside. Crack at 0.08 m from the corner across a length of 0.60 m. Crack corresponding to that observed on the outside. Crack at 0.04 m from the corner starting at 1.10 m from support 9 up to the support.	

Beam C-8/9	Defect no. 1
Outside. One longitudinal crack of length 0.30 m at 0.04 m from the corner and 1.35 m from support 8.	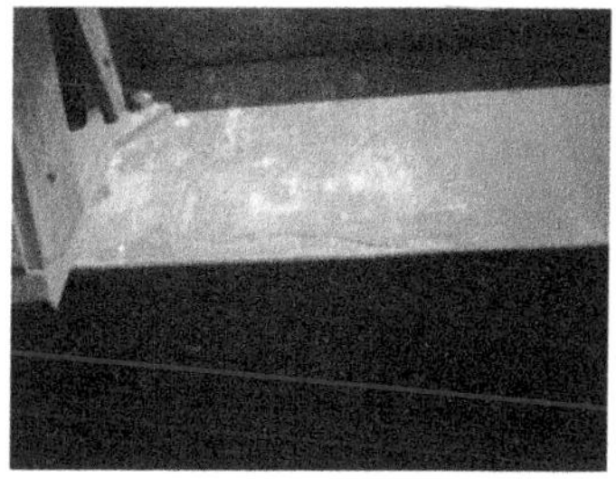

Beam C-8/9	Defect no. 2
Inside. One longitudinal crack at 0.10 m from the underside of the slab over a length of 1.40 m from support 9. One corner crack.	

Beam C-8/9	Defect no. 3
Underside. One crack corresponding to that of the inside.	

Beam C-7/8	Defect no. 1
Outside. One longitudinal crack of length 3.30 m at 0.04 m from the corner and 0.90 m from support 8. One longitudinal crack of length 0.40 m.	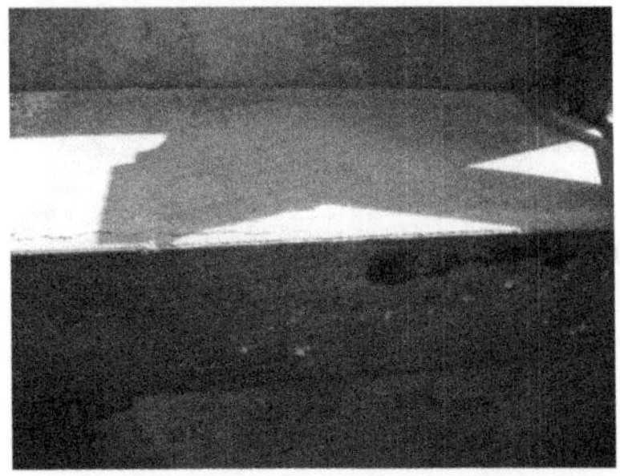

Beam C-7/8	Defect no. 2
Inside. Nothing to report except for a piece of concrete missing 0.80 m from support 8.	

Beam C-7/8	Defect no. 3
Underside. Cracks corresponding to those on the outside. One crack of 0.80 m at 0.06 m from the corner and 2.50 m from support 8. One crack of length 1.80 m at 0.04 m from the corner and 2 m from support 7.	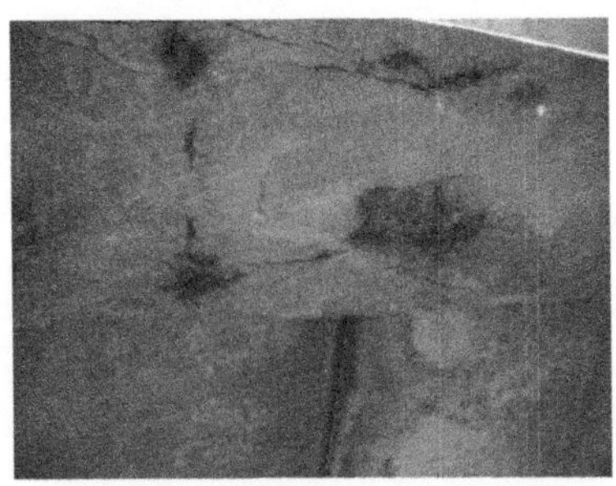

Beam C-5/6	Defect no. 1
Outside. One longitudinal crack of length 0.80 m at 0.10 m from the upper corner. One longitudinal crack in the upper part from the middle of the beam up to support 5.	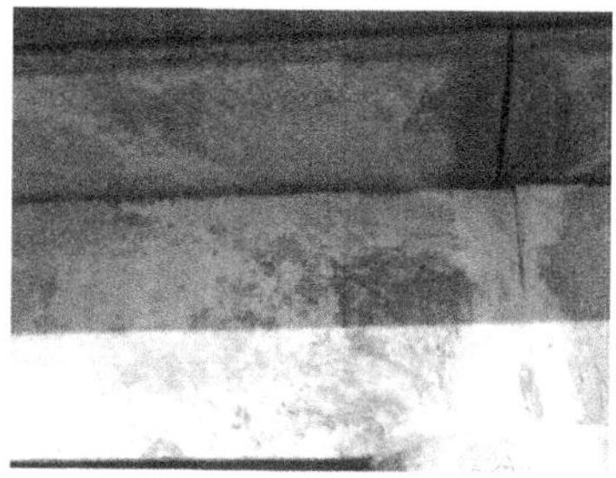

Beam C-5/6	Defect no. 2
Inside. One longitudinal crack in the top part, 1 m long from support 6.	

Beam C-5/6	Defect no. 3
Underside. Cracks corresponding to those of the lateral faces. Highly corroded visible steels. Concrete is absent over length of 2 m. Crack along support 5.	

Beam C-4/5	Defect no. 1
Outside. Nothing to report.	

Beam C-4/5	Defect no. 2
Inside. Crack at 45° to support 5.	

Beam C-4/5	Defect no. 3
Underside. Absence of concrete over length of 3 m at 0.30 m from support 5. Crack at 2.25 m from support 4 over length of 0.40 m. Corrosion of reinforcement.	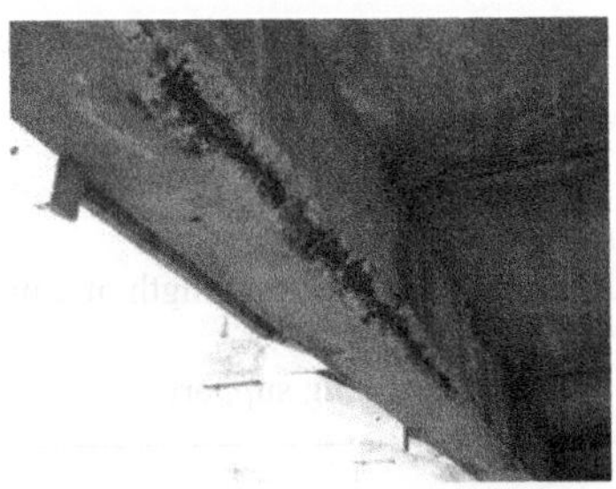

Beam C-3/4	Defect no. 1
Outside. Nothing to report.	

Beam C-3/4	Defect no. 2
Inside. Longitudinal crack in the upper part at the level of support 4 over a length of 1.20 m at 0.10 m from the underside. Renovations were carried out at support 3.	

Beam C-3/4	Defect no. 3
Underside. Support 4.	

Beam D-12/13	Defect no. 1
Exterior is sound except at the level of the wooden console near support 13.	

Beam D-12/13	Defect no. 2
Inside. Longitudinal crack of length 1.80 m at 0.08 m from the lower corner, with a detachment of concrete on the underside. Longitudinal crack at 3 m from support 12 of length 0.50 m. Longitudinal crack at 1.90 m from support 12 up to support 13 at 0.08 m of the corner.	
	Traces of corrosion.

Beam D-12/13	Defect no. 3
Underside. Cracks corresponding to the cracks on the sides.	
	Traces of corrosion.

Beam D-11/12	Defect no. 1
Healthy exterior.	
	No trace of corrosion.

Beam D-11/12	Defect no. 2
Inside. Longitudinal crack of length 0.20 m at 0.80 m from support 12. Longitudinal crack of length 0.50 m at 2.60 m from support 12. Longitudinal crack of length 0.40 m at 1.40 m from support 11.	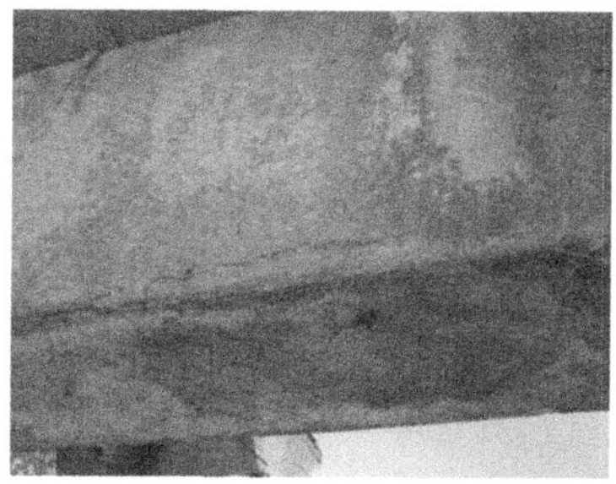
	Slight traces of corrosion.

Beam D-11/12	Defect no. 3
Underside. Cracks corresponding to the cracks on the sides.	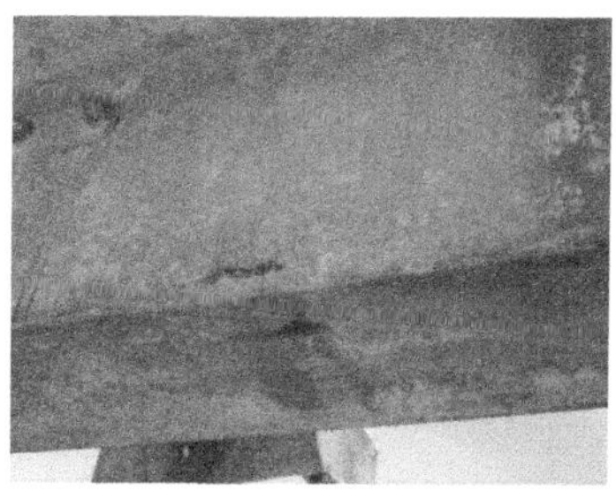
	Traces of corrosion.

Beam D-10/11	Defect no. 1
Healthy outside face.	

Beam D-10/11	Defect no. 2
Inside. Longitudinal crack of length 1.60 m from support 11. Longitudinal crack of length 1.50 m at the middle. Longitudinal crack of length 1.25 m up to support 10.	

Beam D-10/11	Defect no. 3
Underside. Cracks corresponding to the cracks on the sides.	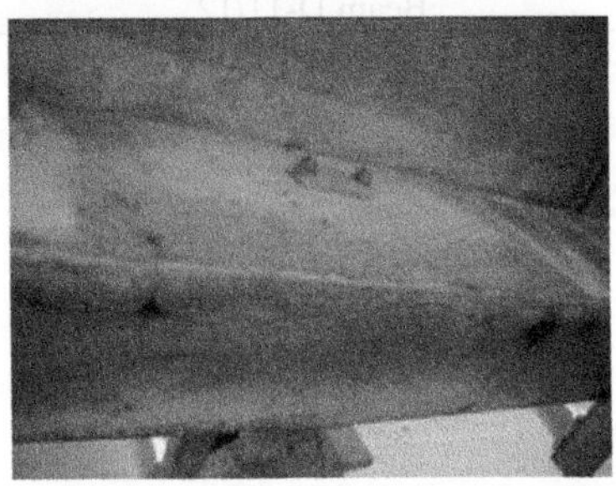

Beam D-9/10	Defect no. 1
Outside. Longitudinal crack in the upper part of length 0.40 m from 1.10 m from support 9.	

Beam D-9/10	Defect no. 2
Outside. Longitudinal crack of length 0.60 m starting at 0.70 m from support 10. Longitudinal crack of length 0.50 m at the middle. Longitudinal crack of length 1.75 m up to support 9.	

Beam D-9/10	Defect no. 3
Underside. Cracks corresponding to the cracks on the side faces.	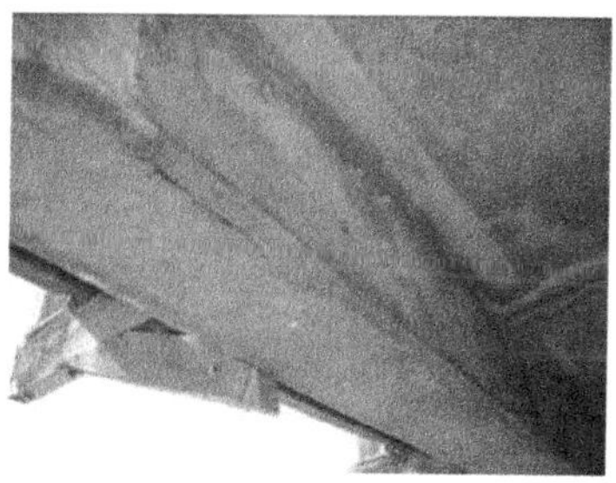

A3.2.1. *Diagrams of Ferroscan auscultations*

Beam D-8/9	Defect no. 1
Outside. Nothing to report.	

Beam D-8/9	Defect no. 2
Inside. Longitudinal crack of length 1 m starting from support 9. Longitudinal crack along support 8: crack at 45°.	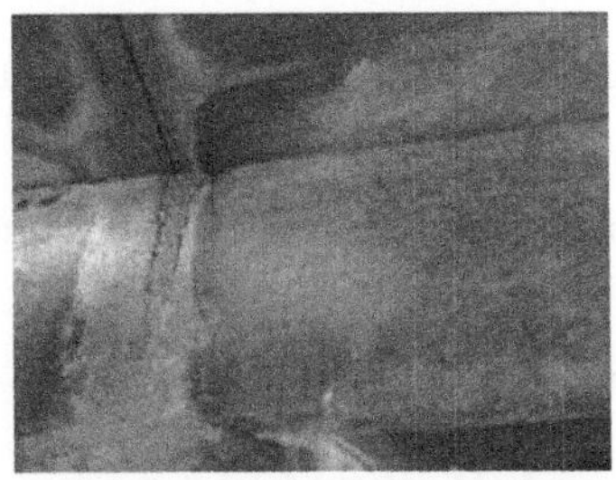

Beam D-8/9	Defect no. 3
Underside. Cracks corresponding to the cracks on the side faces. Concrete deformed by corrosion of reinforcements.	

Beam D-7/8	Defect no. 1
Outside. Nothing to report except for a crack in the middle of length 0.40 m at 0.10 m from the top.	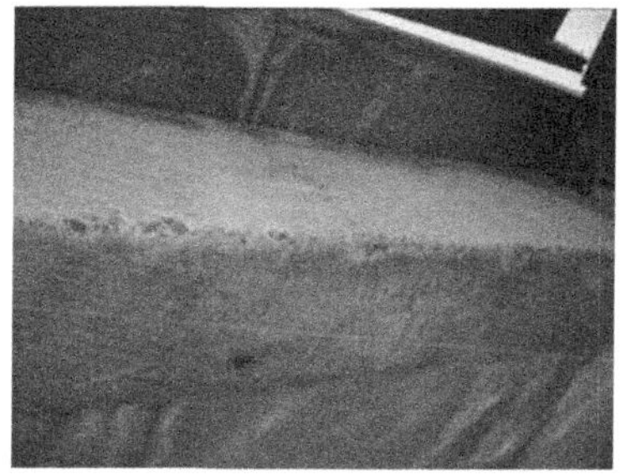

Beam D-7/8	Defect no. 2
Inside. Longitudinal crack from support 8 to mid-span. Longitudinal crack at the console at 0.90 m from support 7.	

Beam D-7/8	Defect no. 3
Underside. Cracks corresponding to the cracks on the sides. Crack at smaller corner to support 8. Crack at 1.30 m from the support 7 over a length of 0.15 m.	

Beam D-6/7	Defect no. 1
Outside. Nothing to report except for segregation at the lower corner.	

Beam D-6/7	Defect no. 2
Inside. Nothing to report.	

Beam D-6/7	Defect no. 3
Underside. One crack of length 0.20m at 1.85 m from support 7.	

Beam D-5/6	Defect no. 1
Outside. Nothing to report.	

Beam D-5/6	Defect no. 2
Inside. Longitudinal crack of length 1.75 m from support 6. Longitudinal crack of length 0.80 m at 1 m from support 5.	

Beam D-5/6	Defect no. 3
Underside. One crack at 0.40 m from support 6. One longitudinal crack of length 1.80 m at middle. Visible steels, detachment of concrete.	

Beam D-4/5	Defect no. 1
Outside. Crack at 45° to support 5. Crack of length 1 m starting at support 5.	

Beam D-4/5	Defect no. 2
Inside. Longitudinal crack of length 0.75 m at 1 m from support 5. Longitudinal crack of length 0.80 m at 1 m from support 5.	

Beam D-4/5	Defect no. 3
Underside. One crack of length 1.20 m in the middle. Longitudinal cracks corresponding to the lateral faces.	

Beam D-3/4	Defect no. 1
Outside. 3 m crack at 0.50 m from support 3.	

Beam D-3/4	Defect no. 2
Inside. Longitudinal crack of length 1 m in the middle. Longitudinal crack of length 1.20 m at 1.55 m from support 3.	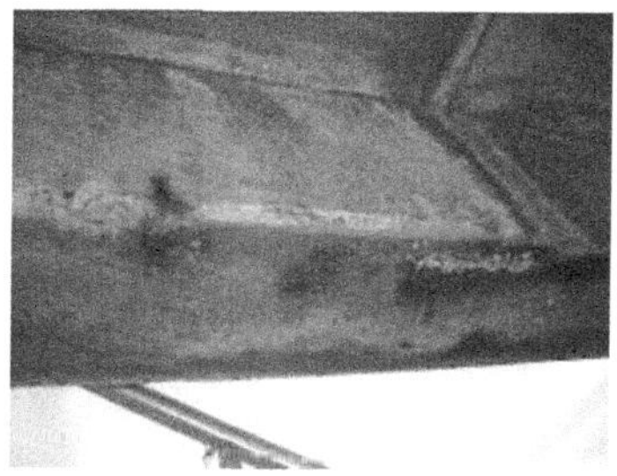

Beam D-3/4	Defect no. 3
Underside. One crack of length 1.20 m at 1.15 m from support 3.	

Beam A1/2/3	Defect no. 1
Outside. Two vertical cracks at 1/3 and ½ span.	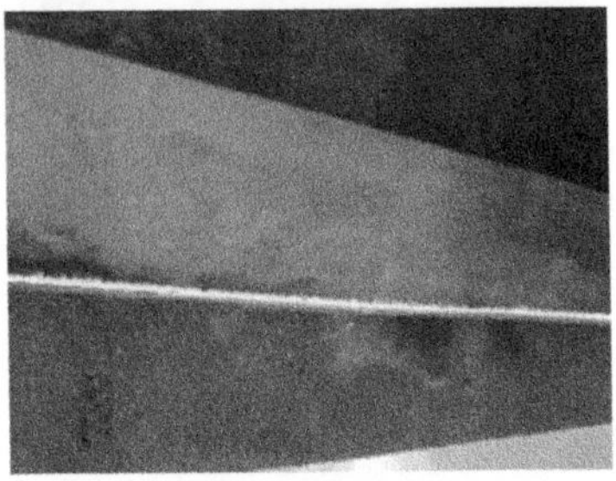

Beam A1/2/3	Defect no. 2
Inside. Nothing to report.	

Beam A1/2/3	Defect no. 3
Underside. Longitudinal crack of length 4.20 m up to support 1.	

Beam B1/2/3	Defect no. 1
Outside. Crack at 1.75 m from support 1 over a length of 0.40 m. Crack of length 0.50 m starting at support 2. Vertical cracks.	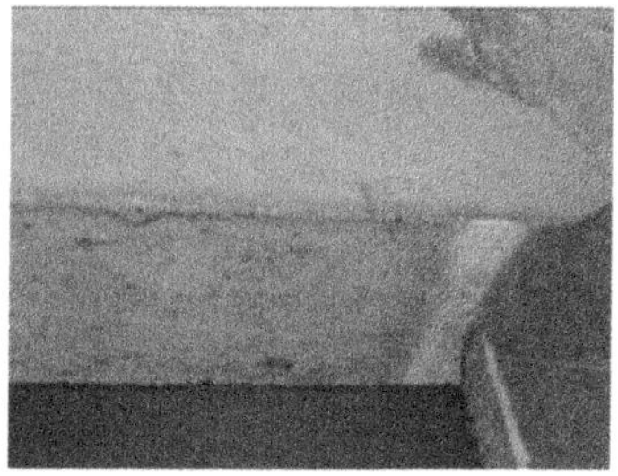

Beam B1/2/3	Defect no. 2
Inside. Vertical cracks identical to previous ones. Vertical crack at 1.85 m from support 2.	

Beam B1/2/3	Defect no. 3
Underside. Longitudinal crack in the middle of length 0.80 m.	

Beam C1/2/3	Defect no. 1
Outside. Nothing to report.	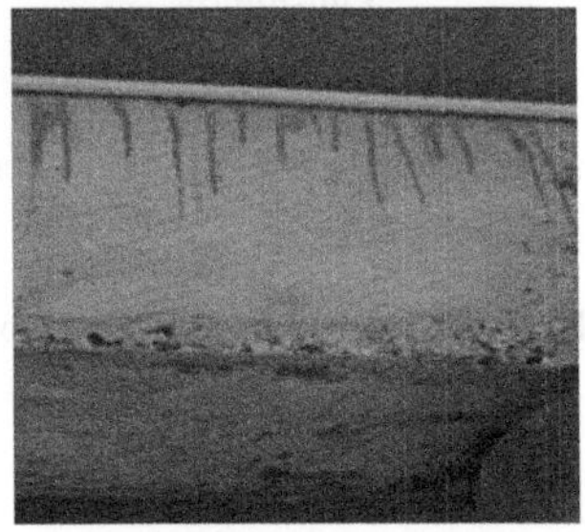

Beam C1/2/3	Defect no. 2
Inside. Nothing to report.	

Beam C1/2/3	Defect no. 3
Underside. Longitudinal crack in the middle.	

Beam D1/2/3	Defect no. 1
Outside. Nothing to report (slight vertical cracking).	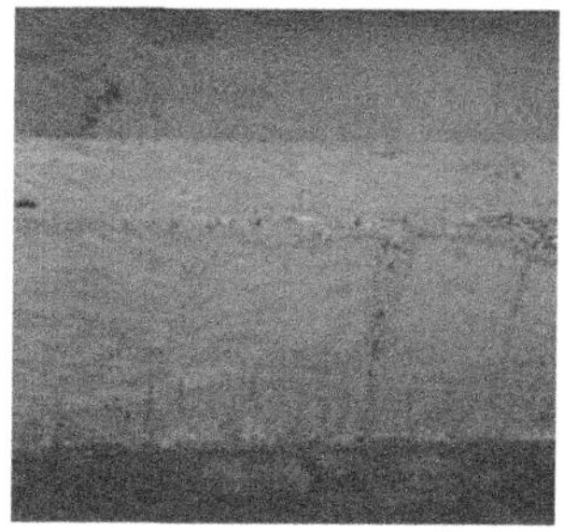

Beam D1/2/3	Defect no. 2
Inside. Nothing to report.	

Beam D1/2/3	Defect no. 3
Underside. Nothing to report.	

Beam E1/2/3	Defect no. 1
Outside. Nothing to report.	

Beam E1/2/3	Defect no. 2
Inside. Crack at support 2 in line with slab BA.	

Beam E1/2/3	Defect no. 3
Underside. Crack in the middle of length 0.30 m.	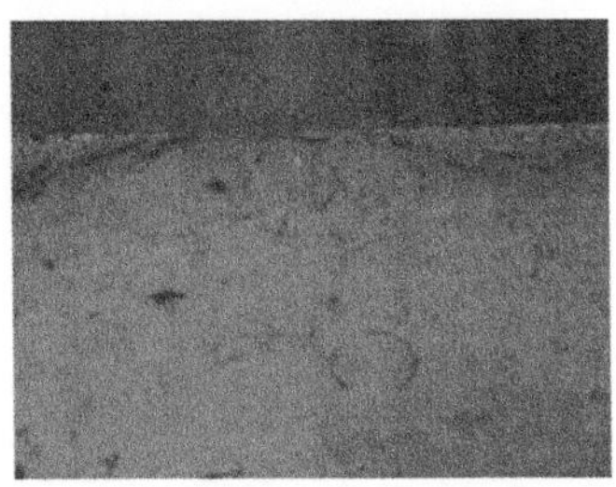

Beam F1/2/3	Defect no. 1
Outside. Vertical crack at mid-span and one-third of span on support side 2. End of the beam cracked.	

Beam F1/2/3	Defect no. 2
Inside. This face has already been renovated.	

Beam F1/2/3	Defect no. 3
Underside. Crack in middle of length 1.50 m.	

A3.2.2. *Directory of pathologies of columns*

Column number	Identified pathology
1A	Vertical crack from top to the mid-height
1B	Two vertical cracks from the top to the mid-height
1C	Segregation of concrete at the top
1D	Segregation of concrete at the top
1E	Horizontal crack at the mid-height Damaged base without any trace of corroded steel
1F	ASR
2A	Slight cracking at the head of the column
3A	Vertical crack at all heights Two horizontal cracks at 1/4 the height
3B	Ditto 2A over 0.20 m
2C	ASR
3C	Corrosion and vertical cracking of 1 m at the head.
2D	Vertical crack of 0.20 m at the head
3D	Two vertical cracks of length 1.30 m
2E	Alkali-reaction phenomenon to be checked
3E	Vertical crack at the head
2F	Vertical crack at the head of length 0.90 m Vertical crack of length 0.50 m
3F	Vertical crack at the head of length 0.40 m
4C	Vertical crack at the head of length 1 m Diametrically opposite vertical crack
4D	Three vertical cracks of length 0.80 m
5 C and 5D	Casting problems
6C	Vertical crack of length 1 m
6D	ASR
7C	Two vertical cracks at 90° of length 1.50 m
7D	ASR
8C	One vertical crack at the head
8D	ASR
9C	Two vertical cracks at the head of length 0.40 m
9D	Verification of alkali reaction Lack of concrete in the lower part

10C	One vertical crack over 0.70 m in the middle
10D	Ditto 10C One vertical crack over 1 m at the top
11C	ASR
11D	ASR
12C	ASR
12D	Lack of concrete at the foot of the column
13C	ASR
13D	ASR

A3.2.3. *Directory of slab pathologies*

Between C and D: Crack in the direction of the impact.	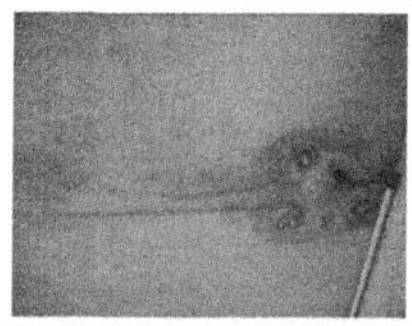
Between D and E: Crack perpendicular to the span of length 5 m.	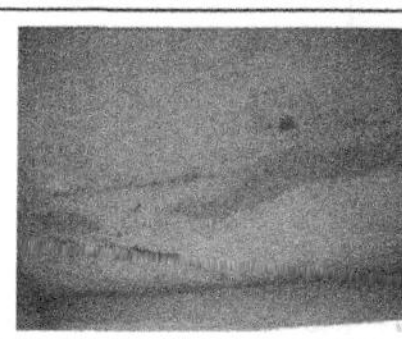
Between E and F: One crack parallel to the span between 1 and 2 One crack parallel to span at 2 One corner crack.	
Between B and C: One diagonal crack between 2 and 3 One crack parallel to span One crack perpendicular to the span.	
Between A and B: One crack parallel to the span at the level of 2.	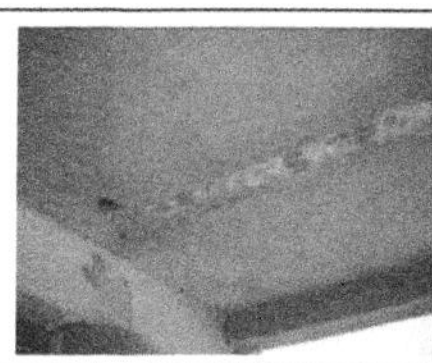

A3.3. Appendix 3c: Sclerometer indices

A3.3.1. *Table of sclerometric indices and related resistance values*

Location	Measured indices	Average value	Compressive strength of concrete on cube (in MPa)	Observations. Value on cylindrical specimen (approximate value in MPa)
Columns				
C4	44/44/46	45	49	40
D4	40/38/34	37	34	27
C5	42/40/40	41	40	38
D5	38/42/42	41	41	40
C6	44/44/42	43	43	40
D6	38/32/40	37	34	27
C7	60/38/42	47	50	45
D7	38/42/40	40	40	38
C8	54/44/48	49	49	38
D8	40/42/32	38	38	34
C9	40/40/42	41	41	40
D9	32/34/32	33	28	18
C10	34/40/30	35	32	23
D10	38/38/40	39	39	37
C11	42/38/44	41	41	40
D11	34/34/40	36	32	23
C12	44/40/44	43	46	40
D12	40/34/40	38	38	34
C13	44/44/48	45	49	40
D13	44/34/38	39	39	35

A1	42/42/46	43	46	40
B1	42/42/40	41	41	40
C1	42/40/42	41	41	40
D1	40/34/32	35	32	23
E1	40/42/36	39	39	37
F1	36/26/32	31	26	15
Beams				
C3 (stretched fiber)	52/43/46	48	60	58
D3 (stretched fiber)	52/54/60	55	65	70
C3 (beam girder)	50/50/50	50	62	59
D3 (beam girder)	52/52/52	52	64	65

NOTE.–

– The other beams were not examined due to the significant deterioration of some of them, which caused the tests to not be representative;

– It should be noted that the results of sclerometric tests are particularly influenced by the following parameters:

- the nature of the cement and its dosage;

- the nature of aggregates;

- the nature of the surface and its surface moisture;

- carbonation: the overestimation of resistance in this case can reach 50%. It is probable that the mechanical strength of the beams is largely overestimated due to the high carbonation of the surface concrete.

A3.4. Appendix 3d: results of pachometric tests on slabs

Quickscan: FQ000155.XFF

Date/Time: 2010-06-04 11:27:45 Bar: 8mm SSN: 10308001

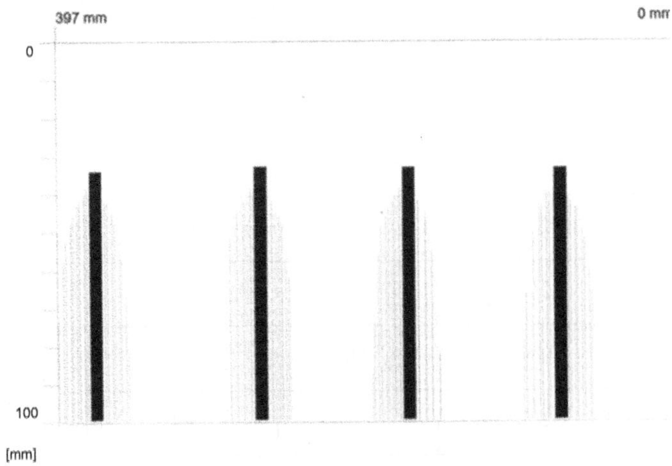

Quickscan Statistics:

Minimum depth:	33 mm	T1:	100 mm
Maximum depth:	34 mm	Number of bars above T1:	4
Bar average:	33 mm	T2:	100 mm
Standard deviation:	1 mm	Number of bars above T2:	4
Cutoff:	100 mm	T3:	100 mm
Number of bars above the cutoff:	4	Number of bars above T3:	4

Client: Mayor

Location: Pier Operator: XL

Comments:

Slab between rows 5 and 6 (underneath).

Archiver le fichier: C:\PROGRA~1\Hilti\PS200\5.3\Download\Prj00002\FQ000155.XFF

A3.4.1. *Results of pachometric tests on the columns*

Quickscan: FQ000045.XFF

Date/Time: 2010-06-03 07:50:02 **Bar:** 6 mm SSN: 10308001

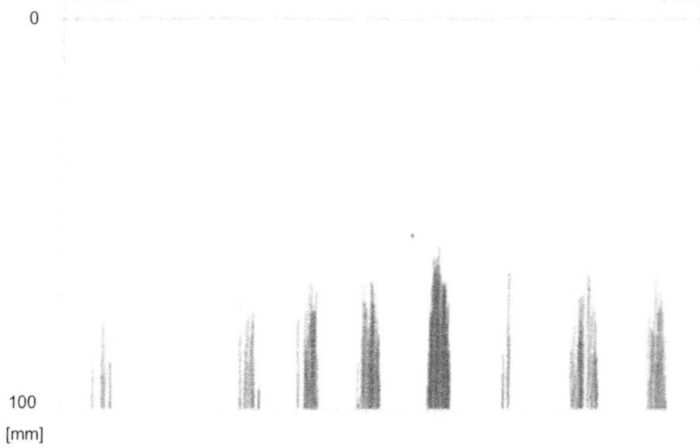

Quickscan Statistics:

Minimum depth:	53 mm	T1:	100 mm
Maximum depth:	93 mm	**Number of bars above T1**	16
Bar average:	73 mm	T2:	100 mm
Standard deviation:	10 mm	**Number of bars above T2**	16
Cutoff:	100 mm	T3:	100 mm
Number of bars above the cutoff:	16	**Number of bars above T3**	16

Client: **Mayor**

Location: **Pier** **Operator:** XL

Comments

Column A1 in vertical direction

A3.4.2. *Results of pachometric tests on the beams*

Quickscan: FQ000070.XFF

Date/Time: 2010-06-03 09:30:14 Bar: 6 mm SSN: 10308001

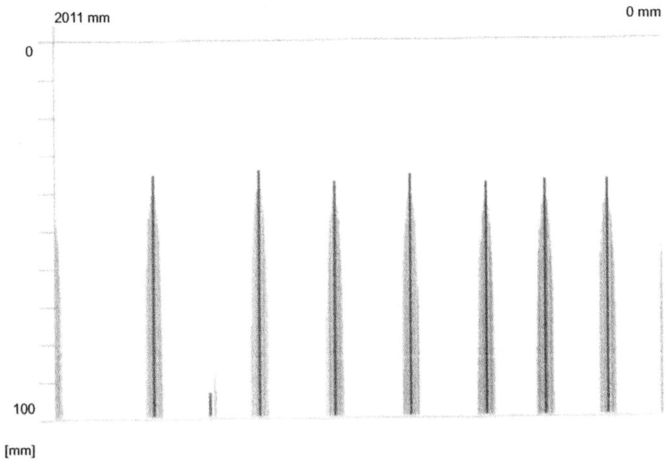

Quickscan Statistics:

Minimum depth:	35 mm	T1:	100 mm
Maximum depth:	94 mm	Number of bars above T1:	8
Bar average:	43 mm	T2:	100 mm
Standard deviation:	20 mm	Number of bars above T2:	8
Cutoff:	100 mm	T3:	100 mm
Number of bars above the cutoff:	8	Number of bars above T3:	8

Client: Mayor

Location: Pier Operator: XL

Comments:

Columns 3-4 row D : lateral face

Appendix 4

Inspection Report "Gantries, Metal Hangers and High Masts"

A4.1. Identification

Identification number:

Manufacturer:

Year of manufacture:

Reference point:

Route:

Direction:

Managing service:

A4.2. General characteristics

A4.2.1. *Solid anchor*

Type of structure:

Materials:

Total length:

Overall width:

A4.2.2. *Gantry*

Fixations on the solid part:

Deck:

Base:

Mount:

Crossbeam:

Device for fixing panels:

Support for messages:

Access devices:

A4.3. Life of the structure

A4.3.1. *Construction of the structure*

Construction date:

Builder:

A4.3.2. *Monitoring action*

Date	Comments
	Not applicable as part of an IDI

A4.3.3. *Maintenance and repair work*

Date	Work done	Location
	Not applicable as part of an IDI	

A4.4. Conditions for access

Date:

Inspection team:

Means used:

Weather conditions

Temperature:

Weather:

Special conditions

Difficulties:

Incidents:

Miscellaneous remarks:

A4.5. General information

A4.5.1. *Purpose of the service*

Apave's mission is to establish a state 0 of the structure described above. This state 0 is carried out within the framework of an initial detailed inspection (IDI) prior to the reception of the work. This IDI must comply with the requirements in Part 2 of booklet 02.

A4.5.2. *Scope of the service*

Inspection and reception missions on civil engineering structures are framed by the technical instruction for the surveillance and maintenance of civil engineering structures (ITSEOA, from the French, *I*nstruction *T*echnique pour la *S*urveillance et l'*E*ntretien des *O*uvrages d'*A*rt, from December 2010 supplemented by the other booklets in the second part of the ITSEOA from 1979).

A4.5.3. *Main references*

Inspections are carried out depending on the type of structure according to:

– the ITSEOA from December 2010 (three booklets);

– the LCPC's Technical Guide "Gantries, metal hangers and high masts";

– the following pamphlets in the second part of the 1979 version of the ITSEOA which is now a technical guide:

- Booklet 02: General information on monitoring;
- Booklet 10: Foundations in aquatic sites;
- Booklet 11: Foundations in land sites;
- Booklet 12: Supports;
- Booklet 13: Support devices;
- Booklet 20: Area impacted, access and approach to structures;
- Booklet 21: Civil engineering structures' equipment;
- Booklet 30: Masonry bridges and viaducts;
- Booklet 31: Bridges of unreinforced and reinforced concrete;
- Booklet 32: Prestressed concrete bridges;
- Booklet 33: Metal bridges;
- Booklet 35: Emergency bridges;
- Booklet 37: Wooden bridges;
- Booklet 40: Tunnels, covered trenches, protective galleries;
- Booklet 50: Metal nozzles;
- Booklet 51: Supporting structures;
- Booklet 52: Excavations and backfill;
- Booklet 53: Protective structures;

– technical instructions issued by SETRA.

COMMENT.– Certain provisions in the special booklets deal with subjects that are identical to the subjects covered in the structural Eurocodes. These are being eliminated from the normative landscape, thus we also integrate the provisions from Eurocodes 2 (concrete structures), 3 (steel structures), 4 mixed structures) and covering, in part 2, the dimensioning of bridges.

A4.6. Annotation of findings

A4.6.1. *Classification of structures*

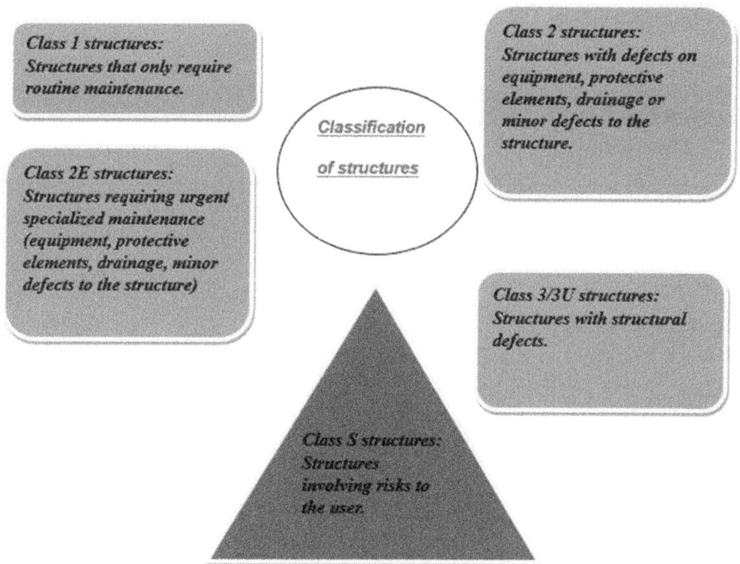

Figure A4.1. *Classification of structures*

A4.7. Gantry

Solid anchor	**Observation sheet no. 001**	
Observations	Category	Photos
Surroundings – Instability of the terrain – Collapsing, slipping, gullying, scouring – Aggressiveness to concrete – Vegetation – Water stagnation		

Concrete quality – Visible mass – Exposed cracks – Surface shrinkage – Visible reinforcements – Corrosion of steels	2	
Base	**Observation sheet no. 002**	
Observations	Category	Photos
Condition of the base – Presence of earth – Water retention – Degraded anti-corrosion protection – *corrosion – Deforming the deck – Deformation of the gussets *Welding conditions* – Cracks – Blow holes – Lack of material – Corrosion – Evolutionary defect – NDC implemented *Fixation to the solid mass* – Presence of a cushioning system – Condition of the blockage – Number of rods – Missing parts, bolts, rods – Loose elements (bolts) – Brake-nut system – Condition of the bolts	2	

Mounting	**Observation sheet no. 003**	
Observations	Category	Photos
General appearance – Geometric defect – Localized deformation – Presence of shocks – Degraded anti-corrosion protection – Corrosion *Verification of welds* – Cracks – Blow holes	2	
– Lack of material – Corrosion *Access hatch* – Presence of a hatch – Presence of closing elements – Waterproofing of the hatch – Condition of the bolts		

Crossbeam	**Observation sheet no. 004**	
Observations	Category	Photos
General appearance – Geometric defect – Localized deformation – Presence of shocks – Degraded anti-corrosion protection – Corrosion – Cross-beam connection *Verification of welds* – Cracks – Blow holes	1	

– Lack of material		
– Corrosion		
Access hatch	2	
– Presence of a hatch		
– Presence of closing elements	5	
– Waterproofing of the hatch		
– Condition of the bolts		

Panel mounting devices		**Observation sheet no. 002**
Observations	Category	Photos
General appearance	1	
– Geometric defect		
– Localized deformation	2	
– Presence of shocks		
– Degraded anticorrosion protection	5	
– Corrosion		
Verification of fixation	2	
– Type of fixation		
– Number of missing elements		
– Number of loose elements		
– Presence of locking nuts	2	
– Corrosion		
Access hatch		
– Presence of a hatch	2	
– Presence of closing elements		
– Waterproofing of the hatch	2	
– Condition of the bolts		

A4.8. Conclusions

A4.8.1. *Summary*

The main damages are as follows:

Equipment

Crossbeam:

Mounting:

Column bases:

Foundations:

A4.8.2 *Reminder on the conclusions from the last management actions of the structure*

Not applicable as part of an IDI.

A4.8.3. *Interpretations of measurements and acknowledgments*

A4.8.4. *Analysis of the causes and significance of damages*

A4.9. Actions to be taken

A4.9.1. *As part of routine maintenance*

A4.9.2. *As part of specialized maintenance*

A4.9.3. *As part of repairs/reinforcements in order of priority*

A4.9.4. *Suggestions for development*

A4.9.5. *Proposals for specific investigations or monitoring*

A4.9.6. *Proposals for security measures*

Appendix 5

Measuring Equipment

Type of information	Nomenclature	Measured variable	Type of control
Concrete compressive strength	Sclerometer	Surface hardness	NDC
	Windsor probe	Surface hardness	NDC
	Crush test on sample	Compressive strength	DC
	Ultrasound	Density variation	NDC
Composition of concrete	Chemical analysis	Concrete components and pathogens	DC
	Mineralogical analysis	Atomic structure	DC
Homogeneity of the concrete	Ultrasound	Density variation	NDC
Position of reinforcements in concrete	Pachometer	Disturbance of the magnetic field by metal	NDC
	High frequency radar	Density variation	NDC
Characterization of steel	Metallographic analysis	Atomic structure	DC
	Tensile test	Elastic limit and stretching	DC
Carbonation of concrete	Chemical analysis	Concrete components and pathogens	DC
	Phenolphthalein	pH	DC
Corrosion condition of a reinforcement in concrete	Measurement of potential	Electrochemical potential difference	NDC
Corrosion rate of a metal reinforcement	Measurement of potential	Electrochemical potential difference	NDC

Verticality	Laser level	Distance	NDC
	Theodolite	Position of a point in space	NDC
Anticorrosion protection	Thickness measurement Sponge electrode potential measurement	Thickness Porosity	NDC
	Thickness (PIG) Measurement of potential at the electric brush	Thickness (2–1,800 µm) Porosity	DC
Cracks in concrete	Fissurometer	Opening and depth of cracks	NDC
Constraints in a material	Strain gauge	Force applied to the volumetric structure	DC
Pretensioning cables	Acoustic study Ultrasound	Cable break Cable integrity	NDC
Welding	Penetration testing	Penetration of liquid	NDC
	Magnetoscopy	Magnetic field	NDC
	Radiography	X-rays	NDC
	Ultrasound	Wave	NDC

NDC, non-destructive control; DC, destructive control.

Table A5.1. *List of measuring equipment*

Appendix 6

Inspections of Bridges

Structures concerned	Aim	Inspection steps	Organization	Operation
All structures must undergo an annual inspection if they are not already involved in a monitoring action within the same year.	– Identify any change in the aforementioned damages in reports from previous visits (Image Quality Structure [IQS] or periodic detailed inspections [PDI]). – Observe the presence of serious damage that has not yet been detected. – Establish a list of maintenance actions to be taken.	It must be possible to see the accessible parts of the structure and to detect any significant changes. A statement listing any serious damage must be established.	– Selection of inspectors. – Establishment of the inspection program. – Planning of maintenance work.	In case of any serious damage detected during the inspection, taking appropriate operating measures into account.
Annual inspection				

Table A6.1. *Details of annual inspection*

Structures concerned	Aim	Inspection steps	Organization	Operation
All routine and non-routine engineering structures undergo a triennial visit.	Determine the class of the structure according to the IQS methodology in order to estimate its evolution relative to its previous classification.	The visiting procedures are defined in the SETRA methodology documents and include: – a visiting guide; – classification of structures. The structures are classified into two lists: – List 1: structures for which there are standard reporting frameworks; – List 2: complex structures without standard reporting.	List the structures to be visited by the IQS by differentiating those from Lists 1 and 2. IQS inspectors should have received training on the IQS methodology. The visit is limited to the accessible parts, the inaccessible parts are subject to a PDI. Except for structures that are specifically not subjected to PDI.	The operation of IQS visiting records makes it possible to establish the classification of structures (see below). The minutes are attached to the project file.

Table A6.2. *Details of triennal visit*

Structures concerned	Aim	Inspection steps	Organization	Operation
Parts of structures that could not be examined during an IQS visit (underwater foundations, submerged supports, supporting structures, etc.). The frequency of specific visits is every 6 years.	Completion of other inspection visits where it was not possible to assess the condition of a structure or major part of a structure.	Ditto IQS.	List of structures subjected to specific visits. The staff responsible for these visits must have the required qualifications (divers, etc.).	The minutes allow: – stopping of the IQS classification of the structure; – completion of the follow-up list to be given to a PDI.
Specific visits				

Table A6.3. *Details of specific visits*

Structures concerned	Aim	Inspection steps	Organization	Operation
All structures must undergo a PDI (routine and non-routine). The frequency of a PDI is every 6 years. For delicate structures, the frequency can be reduced to 3 years or even 1 year. For more robust structures, it can be pushed to 9 years.	Establish a health check report of the structure. This also establishes the IQS class of the structure.	See Appendix 5 of booklet 2, part II.	A plan is established. Inspectors must have received training.	The procedure of the visit report consists of: – possible safeguarding measures (traffic restrictions, etc.); – proposals for enhanced monitoring; – further investigation for diagnosis.
Periodic detailed inspections				

Table A6.4. *Details of periodic detailed inspections*

Structures concerned	Aim	Inspection steps	Organization	Operation
All structures must undergo an initial detailed inspection (IDI) upon their construction as well as after major renovation works (reinforcement of structure, expansion, new management, etc.).	Establish a reference state for subsequent inspections.	The IDI provides a description of the structure from what can be viewed; this implies that the inspection is carried out by competent agents in terms of the construction techniques and the materials used.	See PDI.	It is the reference state of the structure that will allow its evolution to be assessed over time. The report is part of the project file.
Initial detailed inspections				

Table A6.5. *Details of Initial detailed report*

Structures concerned	Aim	Inspection steps	Organization	Operation
All structures, even those that have not undergone a PDI.	To give effect to the guarantees or responsibilities before their expiry.	The structure parts that are subjected to guarantee: – supporting structures; – pavement joints; – waterproofing; – corrosion protection; – etc. Structures at the end of the decennial guarantee	Ditto PDI.	The minutes are to be kept in the project file. They are used for remedial actions.
End-of-contractual warranty visits				

Table A6.6. *Details of End of contractual warranty visits*

Structures concerned	Aim	Inspection steps	Organization	Operation
All structures having suffered damaging phenomena.	Check that the structures have not been damaged by exceptional phenomena (flood, landslide, passage of an exceptional convoy, impact from vehicles, earthquake, act of vandalism, etc.)	Ditto PDI.	The staff must have the same qualifications as for a PDI.	The reports are used like those of a PDI. And kept in the project file.
Exceptional detailed inspections (EDI)				

Table A6.7. *Details of Exceptional detailed inspections*

Bibliography

[CAL 97] CALGARO J-A., LACROIX R., *Maintenance et réparation des ponts,* Presses des Ponts et chaussées, 1997

[ZHA 12] ZHAO J.J., TONIAS E., *Bridge Engineering,* 3rd ed., Mcgraw-Hill Education, 2012.

Annales ITBTP de 1990: "Renforcement et réparation des structures conception et exécution"

Annales ITBTP de 1992: "Réparation et renforcement des structures de bâtiments et d'ouvrages d'art."

CEREMA. Les fascicules de l'ITSEOA.

Code API 653: "Tank Inspection, Repair, Alteration, and Reconstruction"

EN 13670: 2009: Execution of concrete structures

EN 1504-1: Products and systems for the protection and repair of concrete structures. Definitions, requirements, quality control and evaluation of conformity – Part 2: Surface protection systems for concrete

EN 1504-10, Products and systems for the protection and repair of concrete structures. Definitions, requirements, quality control and evaluation of conformity – Part 10 : Site application of products and systems and quality control of the works

EN 1504-3, Products and systems for the protection and repair of concrete structures. Definitions, requirements, quality control and evaluation of conformity – Part 3: Structural and non-structural repair

EN 1504-4, Products and systems for the protection and repair of concrete structures. Definitions, requirements, quality-control and evaluation of conformity – Part 4: Structural bonding

EN 1504-5, Products and systems for the protection and repair of concrete structures. Definitions, requirements, quality control and evaluation of conformity – Part 5: Concrete injection

EN 1504-6, Products and systems for the protection and repair of concrete structures. Definitions, requirements, quality control and evaluation of conformity – Part 6: Anchoring of reinforcing steel bar

EN 1504-7, Products and systems for the protection and repair of concrete structures. Definitions, requirements, quality control and evaluation of conformity – Part 7: Reinforcement corrosion protection

EN 1504-8, Products and systems for the protection and repair of concrete structures. Definitions, requirements, quality control and evaluation of conformity – Part 8: Quality control and evaluation of conformity

ENV 1504-9, Products and systems for the protection and repair of concrete structures. Definitions, requirements, quality control and evaluation of conformity – Part 9: General principles for the use of products and systems

Eurocode 0-EN 1990: Eurocode - Basis of structural design

Eurocode 2: EN 1992-1-1 Design of concrete structures- Part 1-1: General rules for buildings

Fascicule n°6 du STRESS : "La technique de réparation et de renforcement des ouvrages en béton"

Guide de maintenance des ouvrages de génie civil et structures, DT 92 de l'UFIP.

Installations de stockage de céréales. Guide d'inspection et de maintenance. INERIS.

La norme NFP 95-102

Le fascicule n°5 du STRESS

Le fascicule n°6 du STRESS

Les Annales ITBTP n°501 : "Réparation et renforcement de structures de bâtiment et d'ouvrages d'art

Méthode VSC (Visite Simplifiée Comparée). CETMEF.

NF EN 14487-1 Sprayed Concrete- Part 1: Definitions, Specifications and Conformity and NF EN 14487-2 – Sprayed Concrete- Part 2: Execution

NF EN 14488-1 to 7 Testing Sprayed Concrete

NF EN 934-5 Admixtures for Concrete, Mortar and Grout

NFP 95 104: 1992 Engineering Structures- Repair and Strengthening of Concrete and Masonry Construction- Specifications Pertaining to the Prestressing Technique

NFP 95-106: Engineering structures- Repair and Strengthening of Concrete and Masonry Constructions- Specifications Pertaining to the Structure Foundations

Norme EN 12-696 : Protection des aciers dans le béton.

Norme NFP 95-105(en préparation).

Pathologie des ouvrages en béton. Annales ITBTP n° 536.

Pathologie et réparation des ouvrages en béton de stockage et de transport des liquides. Annales ITBTP n°532 March/April 1995.

Portiques, potences, hauts-mâts. Guide de maintenance du LCPC.

Réaction Sulfatique Interne. Guide LCPC de 2007.

Index

Other titles from

ISTE

in

Civil Engineering and Geomechanics

2016

BARRE Francis, BISCH Philippe, CHAUVEL Danièle *et al.*
Control of Cracking in Reinforced Concrete Structures: Research Project CEOS.fr

PIJAUDIER-CABOT Gilles, LA BORDERIE Christian, REESS Thierry, CHEN Wen, MAUREL Olivier, REY-BETHBEDER Franck, DE FERRON Antoine
Electrohydraulic Fracturing of Rocks

TORRENTI Jean-Michel, LA TORRE Francesca
Materials and Infrastructures 1
Materials and Infrastructures 2

2015

AÏT-MOKHTAR Abdelkarim, MILLET Olivier
Structure Design and Degradation Mechanisms in Coastal Environments

MONNET Jacques
In Situ Tests in Geotechnical Engineering

2014

DAÏAN Jean-François
Equilibrium and Transfer in Porous Media – 3-volume series
Equilibrium States – Volume 1
Transfer Laws – Volume 2
Applications, Isothermal Transport, Coupled Transfers – Volume 3

2013

AMZIANE Sofiane, ARNAUD Laurent
Bio-aggregate-based Building Materials: Applications to Hemp Concretes

BONELLI Stéphane
Erosion in Geomechanics Applied to Dams and Levees

CASANDJIAN Charles, CHALLAMEL Noël, LANOS Christophe, HELLESLAND Jostein
Reinforced Concrete Beams, Columns and Frames: Mechanics and Design

GUÉGUEN Philippe
Seismic Vulnerability of Structures

HELLESLAND Jostein, CHALLAMEL Noël, CASANDJIAN Charles, LANOS Christophe
Reinforced Concrete Beams, Columns and Frames: Section and Slender Member Analysis

LALOUI Lyesse, DI DONNA Alice
Energy Geostructures: Innovation in Underground Engineering

LEGCHENKO Anatoly
Magnetic Resonance Imaging for Groundwater

2012

BONELLI Stéphane
Erosion of Geomaterials

JACOB Bernard *et al.*
ICWIM6 – Proceedings of the International Conference on Weigh-In-Motion

OLLIVIER Jean-Pierre, TORRENTI Jean-Marc, CARCASSES Myriam
Physical Properties of Concrete and Concrete Constituents

PIJAUDIER-CABOT Gilles, PEREIRA Jean-Michel
Geomechanics in CO_2 Storage Facilities

2011

BAROTH Julien, BREYSSE Denys, SCHOEFS Franck
Construction Reliability: Safety, Variability and Sustainability

CREMONA Christian
Structural Performance: Probability-based Assessment

HICHER Pierre-Yves
Multiscales Geomechanics: From Soil to Engineering Projects

IONESCU Ioan R. *et al.*
Plasticity of Crystalline Materials: from Dislocations to Continuum

LOUKILI Ahmed
Self Compacting Concrete

MOUTON Yves
Organic Materials for Sustainable Construction

NICOT François, LAMBERT Stéphane
Rockfall Engineering

PENSE-LHERITIER Anne-Marie
Formulation

PIJAUDIER-CABOT Gilles, DUFOUR Frédéric
Damage Mechanics of Cementitious Materials and Structures

RADJAI Farhang, DUBOIS Frédéric
Discrete-element Modeling of Granular Materials

RESPLENDINO Jacques, TOUTLEMONDE François
Designing and Building with UHPFRC

2010

ALSHIBLI A. Khalid
Advances in Computed Tomography for Geomechanics

BUZAUD Eric, IONESCU Ioan R., VOYIADJIS Georges
Materials under Extreme Loadings / Application to Penetration and Impact

LALOUI Lyesse
Mechanics of Unsaturated Geomechanics

NOVA Roberto
Soil Mechanics

SCHREFLER Bernard, DELAGE Pierre
Environmental Geomechanics

TORRENTI Jean-Michel, REYNOUARD Jean-Marie, PIJAUDIER-CABOT Gilles
Mechanical Behavior of Concrete

2009

AURIAULT Jean-Louis, BOUTIN Claude, GEINDREAU Christian
Homogenization of Coupled Phenomena in Heterogenous Media

CAMBOU Bernard, JEAN Michel, RADJAI Fahrang
Micromechanics of Granular Materials

MAZARS Jacky, MILLARD Alain
Dynamic Behavior of Concrete and Seismic Engineering

NICOT François, WAN Richard
Micromechanics of Failure in Granular Geomechanics

2008

BETBEDER-MATIBET Jacques
Seismic Engineering

CAZACU Oana
Multiscale Modeling of Heterogenous Materials

HICHER Pierre-Yves, SHAO Jian-Fu
Soil and Rock Elastoplasticity

JACOB Bernard *et al.*
HVTT 10

JACOB Bernard *et al.*
ICWIM 5

SHAO Jian-Fu, BURLION Nicolas
GeoProc2008

2006

BALAGEAS Daniel, FRITZEN Claus-Peter, GÜEMES Alfredo
Structural Health Monitoring

DESRUES Jacques *et al.*
Advances in X-ray Tomography for Geomaterials

FSTT
Microtunneling and Horizontal Drilling

MOUTON Yves
Organic Materials in Civil Engineering

2005

PIJAUDIER-CABOT Gilles, GÉRARD Bruno, ACKER Paul
Creep Shrinkage and Durability of Concrete and Concrete Structures CONCREEP – 7

Printed and bound by CPI Group (UK) Ltd, Croydon, CR0 4YY
16/09/2026
14963072-0006